<u>Disclaimer</u>

The publisher of this book is by no way associated with the National Institute of Standards and Technology (NIST). The NIST did not publish this book. It was published by 50 page publications under the public domain license.

50 Page Publications.

Book Title: High-temperature tensile constitutive data and models for structural steels in fire

Book Author: William E. Luecke; Stephen W. Banovic; Joseph D. McColskey;

Book Abstract: This report documents the stress-strain behavior of a collection of structural steels recovered from the collapse of the World Trade Center. These steels, combined with literature data form the basis of a model for the stress-strain behavior of structural steels in general. The model accounts for the lost of strength, the decrease in work hardening and the increase in the strain-rate sensitivity with increasing temperature. For general structural steels, it takes the measured yield strength as its only input parameter. The new model predicts the stress-strain behavior of the steels slightly better than the existing Eurocode~3 stress-strain model.

Citation: NIST TN - 1714

Keywords: steel; World Trade Center; tensile, strain rate; constitutive mode

NIST Technical Note 1714

High-temperature tensile constitutive data and models for structural steels in fire

William Luecke
Stephen W. Banovic
J. David McColskey

NIST National Institute of Standards and Technology • U.S. Department of Commerce

NIST Technical Note 1714

High-temperature tensile constitutive data and models for structural steels in fire

William E. Luecke
Stephen W. Banovic
Metallurgy Division
Material Measurement Laboratory

J. David McColskey
Materials Reliability Division
Material Measurement Laboratory

Nov 2011

U.S. Department of Commerce
John E. Bryson, Secretary

National Institute of Standards and Technology
Patrick D. Gallagher, Director

Certain commercial entities, equipment, or materials may be identified in this document in order to describe an experimental procedure or concept adequately. Such identification is not intended to imply recommendation or endorsement by the National Institute of Standards and Technology, nor is it intended to imply that the entities, materials, or equipment are necessarily the best available for the purpose.

National Institute of Standards and Technology Technical Note 1714
Natl. Inst. Stand. Technol. Technical Note 1714, 135 pages (Nov 2011)
CODEN: NTNOEF

Abstract

This report documents a model to represent the true stress-strain, $\sigma - \epsilon$, behavior of structural steel. It is based on combination of data from the NIST World Trade Center collapse investigation and many other evaluated literature sources. Unlike other models for stress-strain behavior of structural steel, such as the Eurocode 3 formulation [1], the model explicitly describes the time-dependent nature of the strength of steel at high temperature. For untested steels, it predicts the stress-strain behavior using only the measured room-temperature yield strength, S_y. The relative deviation between the model of this report and the actual data for the steels is generally less than 25 %, and is always less than 50 %. On subset of eight steels, the model predicts the stress-strain behavior slightly better than the equally complicated Eurocode 3 model. For three literature structural steels, not analyzed as part of the model, the model of this report and the Eurocode 3 model predict stress-strain behavior with similar quality.

Keywords: steel, World Trade Center, tensile, strain rate, constitutive model

Contents

1 Introduction

Modeling the response of steel structures to fire requires accurate constitutive models for the behavior of the steels in the beams, columns, and connections. Finite-element or other analytical models require the entire stress-strain curve, rather than just the summary data like yield and tensile strengths, which are common in the technical literature, for example the ASTM data series [2] that dates back to the 1950s. Unfortunately, some high-quality data sets that form the basis of high-temperature deformation models were produced as technical committee reports [3, 4] or as internal industry reports [5, 6] that were not widely circulated and are difficult to obtain.

For structural steels in conditions relevant to fire, the majority of effort to develop full constitutive laws began in the 1980s in Europe [3, 7–9] and culminated in the stress-strain model of the Eurocode 3 [1]. Even so, full high-temperature stress-strain data sets for structural steels used in buildings are much less common than summary data. The National Institute of Standards and Technology, as part of its report on the collapse of the World Trade Center, characterized many important steels recovered from the buildings to provide stress-strain models to analyze the impact, fires, and resulting collapse. Those tests represent a large additional data set that can be used for modeling the response of steel structures to fire. The nine steels of the present report represent a selection of the steel most likely to have been involved in the fires in the World Trade Center.

This report has four goals.

1. Summarize and reanalyze the high-temperature tensile behavior of steels recovered from World Trade Center collapse.

2. Evaluate the temperature dependence of their strength, stress-strain behavior, and strain rate sensitivities to provide useful constitutive data for modelers who need data for structural steel.

3. Present a method for modeling the high-temperature stress-strain behavior of structural steel that is suitable for determining the response of steel structures to fire.

4. Compare the high-temperature retained strength and the stress-strain model to the Eurocode 3 [1] recommended values.

2 Experimental Methods

The steels were characterized for chemistry and microstructure using standard metallographic techniques. Room-temperature tensile testing at room temperature followed ASTM E8 [10], and high-temperature tensile testing followed ASTM E21

Table 1: Summary of the nine steels in this report.

Abbreviation	Specimen	F_y ksi	Description of shape
C65	C65-f1-1	36	12WF161 flange $t = 37.7$ mm
C80	C80-f1-1	36	14WF184 flange $t = 35$ mm
C128	C128-T1	36	core truss seat channel
HH	HH-f1-1	42	12WF92 flange $t = 22$ mm
C53BA	C53-ba3	36	76 mm × 50 mm angle $t = 9.4$ mm
C132	C132-ta-3	50	50 mm × 28 mm angle $t = 6.35$ mm
C40	C40-c2m-iw-1	60	$t = 6.35$ mm plate
N8	N8-c1b-f-1	60	$t = 7.9$ mm plate
C10	C10-c1m-fl-1	100	$t = 6.35$ mm plate

Notes: F_y : specified yield strength see Appendix A

[11]. Appendix B summarizes the experimental methods in greater detail. Appendix A summarizes the symbols used in this report. The data reported here are abstracted from the NIST World Trade Center collapse investigation report on structural steel properties. [12].

3 Results

3.1 Microstructure and Chemistry of the steels

The nine steels, summarized in Table 1, represent a cross-section of the steels that were important to modeling the fire response of the World Trade Center. To facilitate cross-reference with the original report [12], this report retains the same specimen nomenclature. The set comprises specimens from $F_y = 36$ ksi hot-rolled wide-flange shapes, $F_y = 60$ ksi hot-rolled plates, a quenched-and-tempered $F_y = 100$ ksi plate, and hot-rolled bar and angle taken from the floor trusses. The lower-strength steels were generally supplied to ASTM A36 and ASTM A242. The higher-strength plates were generally supplied to proprietary Japanese specifications, but are similar to contemporaneous U.S. specifications. Appendix C summarizes the sources of the nine steels and their chemical and microstructural characterization in greater detail.

3.2 Tensile deformation

Figure 1 summarizes the true stress-strain behavior of the nine steels. In Figure 1, the true stresses are normalized to the room-temperature $\epsilon = 0.002$ offset yield strength, S_y(0.002 offset), which are listed in Table 6 in Appendix D.1. Figure 1 omits several true stress-strain curves for clarity, but Table 6 contains entries for them. Appendix D.2 plots the un-normalized curves.

4 Discussion

The goal of this report is to produce a model of the stress-strain behavior of structural steel that can predict the behavior of generic structural steels. It is based on as much publicly available, critically evaluated data as possible, to ensure that it represents the behavior of all structural steels, rather than just a few from a single manufacturer or region. It breaks the behavior into three parts:

1. the general behavior of the retained yield strength as a function of temperature, which includes all the available literature data;

2. the behavior of the strain hardening post-yield, which is based on the behavior of the steels evaluated as part of the NIST World Trade Center collapse investigation [12];

3. the behavior of the significant sensitivity of the strength of steel to the deformation rate at elevated temperature, which includes all the available literature data.

Basing the behavior on the retained yield strength allows us to include many more steels than would be possible if the model were limited to literature reports that included full stress-strain behavior.

4.1 Retained Yield Strength

A common method to represent the high-temperature strength of steel is to plot the retained yield or tensile strength as a function of temperature, normalized to the room-temperature value [1, 2, 9, 13–23]. A commonly voiced but undocumented rule of thumb is that the yield strength of structural steel drops to one half its room-temperature value at 538 °C (1000 F). Figure 2 plots retained-strength data from this study, as well as from many literature sources [6, 12, 18, 23–35] , which are summarized in Appendix E.1. The data in the figure and in the analysis are limited to tests whose strain rate lies within or just slightly outside the ranges specified in ISO [36], JIS [37], and ASTM [11] high-temperature tensile test standards. The spread of the data demonstrates the variability in the response of structural steel to elevated temperature deformation.

The shape of the decrease in retained yield strength with temperature is complicated. The NIST WTC collapse report [12] represented it with a five-parameter phenomenological fit where the parameters were constrained and adjusted to visually provide the best representation of the data in the temperature region, $(400 <$

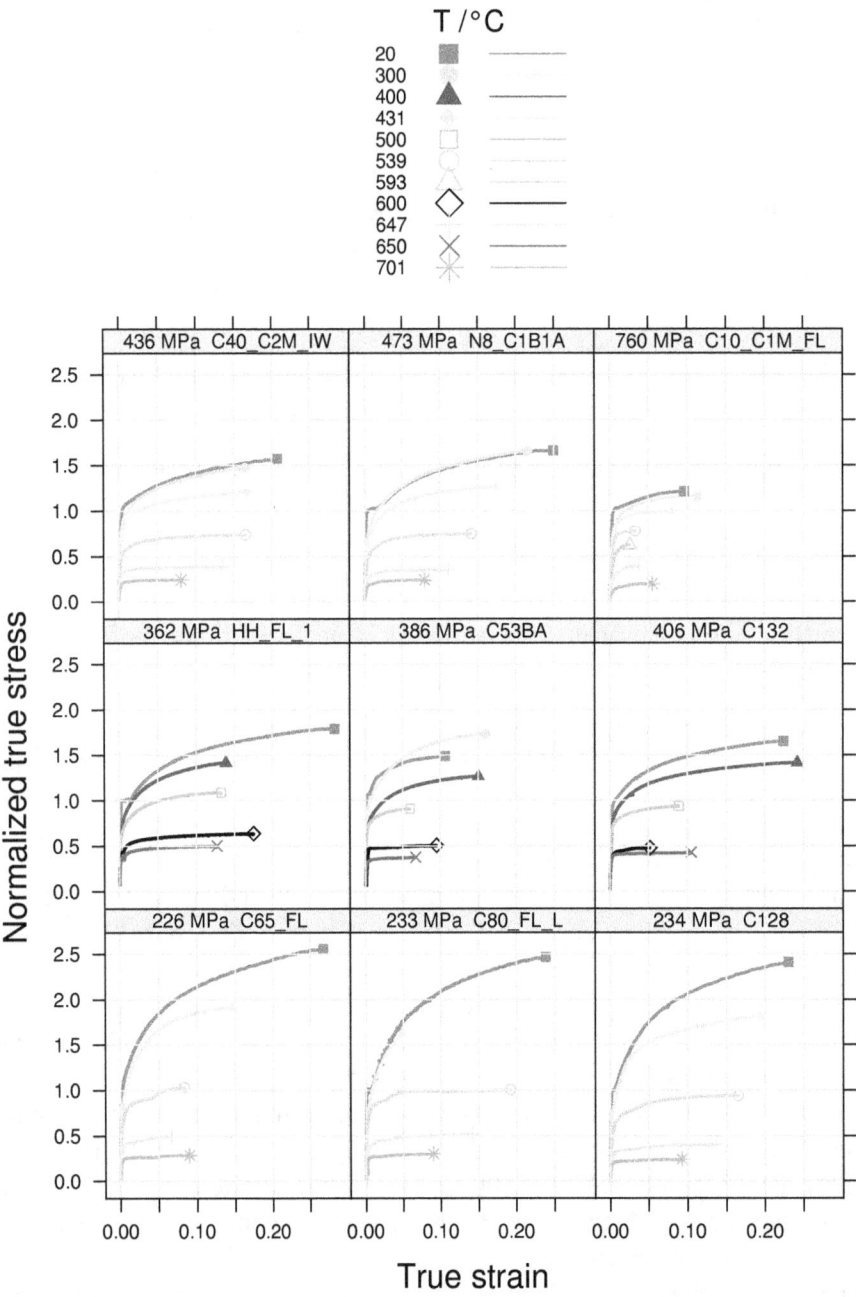

Figure 1: True stress-strain tensile curves for all steels, where stresses are normalized to the measured room-temperature yield strength, $S_y(0.002$ offset), shown in the strip for each plot.

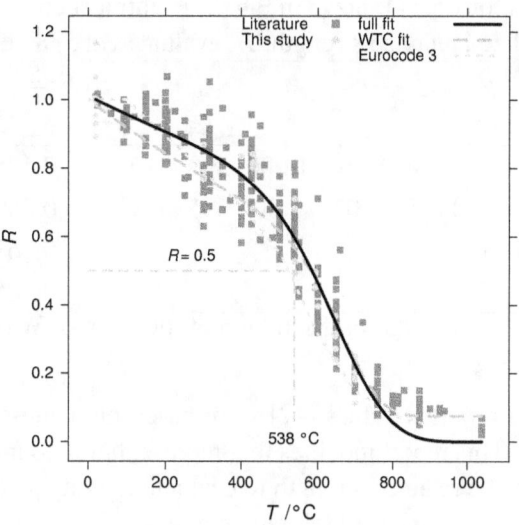

(a) full temperature range

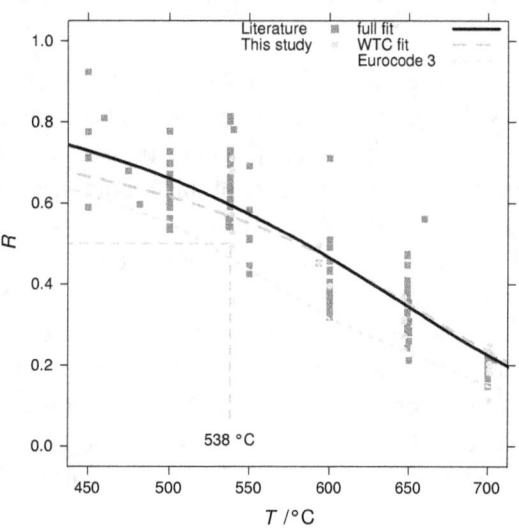

(b) temperature range where strength drop is largest

Figure 2: Normalized, retained yield strength, R, as a function of temperature. The strength is the 0.2 % offset yield strength, S_y(0.002 offset) normalized to its room-temperature value. Data are from structural steels tested in the strain-rate range $(3.30 \times 10^{-5} < \dot{\epsilon} < 1.35 \times 10^{-4})$ 1/s. (a) full temperature range. (b) temperature range where strength drop is largest.

Table 2: Parameters for Eq. (1) shown in Fig. 2. Rightmost column is the predicted value of the normalized retained strength, R, evaluated at 538 °C (1000 F).

Fit	A_2	r_1	r_2	r_3	r_4	$R(T = 538°C)$
				°C	°C	
Full		5.708	1.000	590	919	0.595
WTC	0.075	8.07	1.00	635	539	0.568
Eurocode 3						0.461
ECCS						0.405

See Appendix D.3 for discussion of the form of WTC fit.

$T < 650)$ °C. The reported values [12] were based on a subset of the literature data available here. This report modifies that representation to include more literature data, to require the retained strength to evaluate to unity at $T = 20$ °C, and to asymptotically approach zero for high temperatures. The retained yield strength, R, depends on temperature through a four-parameter model:

$$R = \frac{S_y}{S_y(T = 20°C)} = \exp\left(-\frac{1}{2}\left(\frac{T^*}{r_3}\right)^{r_1} - \frac{1}{2}\left(\frac{T^*}{r_4}\right)^{r_2}\right) \qquad (1)$$

where $T^* = T - 20$, measured in °C. Figure 2 plots three representations of Eq. (1). The curve labeled "WTC fit" uses the parameters that appeared in Ref. 12. The curve labeled "full fit" was computed using a constrained, non-linear least-squares fit to the literature data combined with data of this report, see Appendix E.1, from tests with strain rate $(3.30 \times 10^{-5} < \dot{\epsilon} < 1.35 \times 10^{-4})$ 1/s. Figure 2 also shows a curve of retained strength computed for the Eurocode 3 model [1]; details are in Appendix F.2. The European Convention on Constructional Steelwork [9] also recommended values for retained strength, described in Appendix G. Table 2 summarizes the parameters for both fits to the data. The last column of Table 2 shows the predicted values of R for $T = 538$ °C (1000 F). Both the original prediction from the NIST WTC report [12] and the new prediction are slightly higher than the commonly accepted value $R = 0.5$, while the predictions of other models are lower.

4.2 Modeling stress-strain curves

Analyses often require more than simply the behavior of the yield strength as a function of temperature. Finite-element analyses require the entire stress-strain curve, rather than just an estimate of the yield strength. The Eurocode 3 [1] model uses a multi-parameter model evaluated at fixed temperature points to describe the stress-strain curve, see Appendix F, but does not document the origins of the choices of the parameters. This report presents an alternative method to describe the high-temperature steel stress-strain behavior and describes the method and data by which the parameters were calculated.

Dozens of stress-strain-strain-rate models of different complexity exist. In preparing this report we examined and evaluated a number of them and tried to balance the fidelity of the model against the number of parameters and the ease of calculation. The stress-strain model in this report uses a power-law (or Hollomon) strain-hardening, and a simple power-low strain rate sensitivity that scales the entire true stress-strain curve:

$$\sigma = \left[S_y(T) + K(S_y^0, T)\epsilon_p^n \right] f(\dot{\epsilon}) \tag{2}$$

Equation (2) can be broken into several independent terms.

- $S_y(T)$ is the behavior of the yield strength as a function of temperature, T.

- ϵ_p is the plastic true strain, as opposed to the total true strain.

- $K(S_y^0, T)$ is the behavior of the so-called strength coefficient, as a function of temperature and room-temperature yield strength, S_y^0.

- Together $K\epsilon_p^n$ form the familiar power-law stress-strain equation.

- $f(\dot{\epsilon})$ is the function that describes the strain-rate sensitivity of the stress-strain behavior.

For reasons of familiarity, and to create a model that was similar across all temperatures, elastic behavior is included up to the yield strength at all temperatures. Each stress-strain curve has an elastic region up to the yield strength and a plastic region post-yield.

$$\sigma = \begin{cases} E\epsilon & \text{for } \epsilon < S_y/E \\ S_y + K(\epsilon - S_y/E)^n & \text{for } \epsilon \geq S_y/E \end{cases} \tag{3}$$

See also Figure 3. The total strain, ϵ, is the sum of the elastic, ϵ_e, and plastic, ϵ_p, strains:

$$\epsilon = \epsilon_e + \epsilon_p = \frac{\sigma}{E} + \epsilon_p \tag{4}$$

The following sections demonstrate the functional dependence and computation of the individual terms.

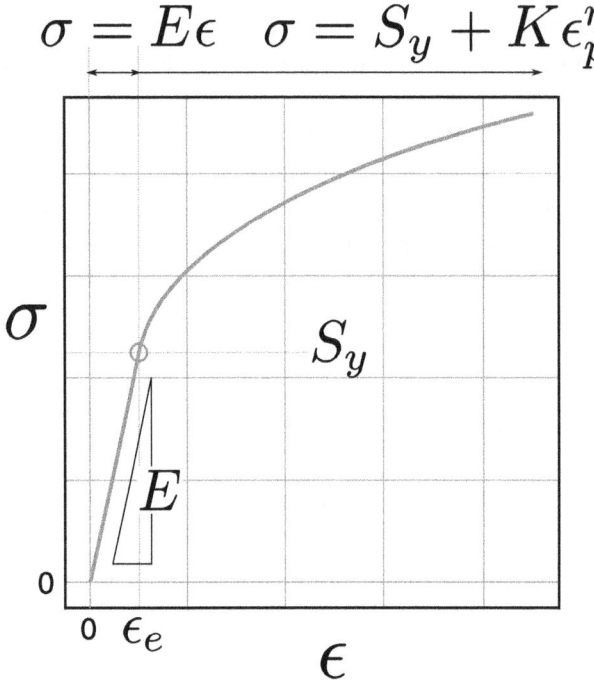

Figure 3: Schematic diagram of the stress-strain model used in this report.

4.2.1 Yield strength

Most of the literature on high-temperature strength of structural steel reports only the yield and tensile strength, and not the full stress-strain behavior. Those yield-strength data were used in computing the retained strength behavior of the previous section: Eq. (1) and Figure 2. Greater fidelity to actual stress-strain behavior can be achieved by incorporating that large body of literature data, Figure 2, on the behavior of the yield strength of structural steel. In Eq. (2) the high-temperature yield strength is

$$S_y = RS_y^0 \tag{5}$$

One limitation of this approach is that the reported yield strength, $S_y(0.2\,\%$ offset$)$ does not occur in the stress-strain curve at the point where stress deviates from a linear increase with strain. However, the small difference that using the yield strength as the end of the elastic region creates is outweighed by the utility of using the large body of literature data to set the yield behavior of the stress-strain curve.

14

4.2.2 Elastic Modulus

The temperature dependence of the elastic modulus is taken from the NIST World Trade Center Collapse report [12]:

$$E = E_0 + e_1 T + e_2 T^2 + e_3 T^3 \tag{6}$$

Table 3 summarizes the values of the parameters. Appendix E.4 discusses the sources and limitations of elastic modulus data and expressions.

4.2.3 Modeling the strain hardening

The stress-strain behavior, Figure 1, demonstrates the general relation that steels with high yield strength generally strain harden less than steels with low yield strength. Figure 4a demonstrates this observation by plotting the ratio of the stress evaluated at $\epsilon = 0.075$ to the yield strength, S_y, as a function of the measured room-temperature yield strength, S_y^0 . The four panels show the data in different temperature ranges. The effect is strong at temperatures up to $450\,^{\circ}\mathrm{C}$, but is weaker at higher temperatures, where time-dependent effects are more significant. The stress-strain model should capture this behavior.

In addition to the slight dependence on the room-temperature yield strength, the amount of strain hardening decreases with increasing temperature for most steels. Figure 4b plots the ratio of the stress evaluated at $\epsilon = 0.075$ to the yield strength, as a function of temperature. The panels separate the steels into three groups by room-temperature yield strength, S_y^0. This ratio, which represents the amount of strain hardening after yield, drops off with temperature for all but the two highest-strength steels.

A model for stress as a function of plastic strain that captures the two requirements above, but that does not use too many parameters, is

$$\sigma = \begin{cases} E\epsilon & \text{for } \epsilon < S_y/E \\ RS_y^0 + (k_3 - k_4 S_y^0) \exp\left(-(\frac{T}{k_2})^{k_1}\right)(\epsilon - S_y/E)^n & \text{for } \epsilon \geq S_y/E \end{cases} \tag{7}$$

The strength coefficient, K in Eq. (2), depends linearly on the room-temperature yield strength. In addition, it drops off with increasing temperature through the extra exponential term. That term causes the amount of strain hardening decrease with increasing temperature, as seen in Figure 4b. The retained yield strength, R, is computed from Eq. (1).

4.2.4 Including strain-rate sensitivity

At elevated temperature the strength of steel measured in a tension test increases significantly as the strain rate at which it is deformed increases. At temperatures

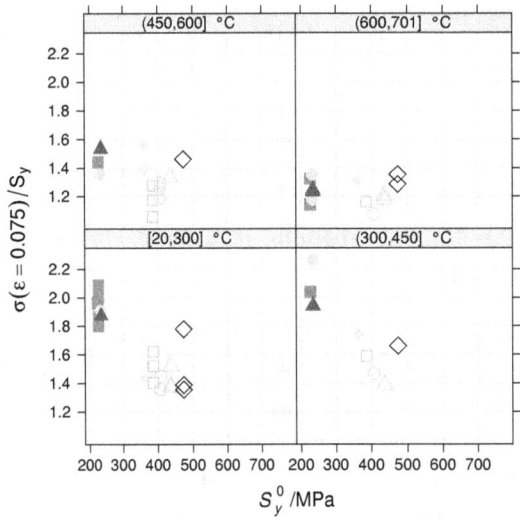

(a) in different temperature ranges

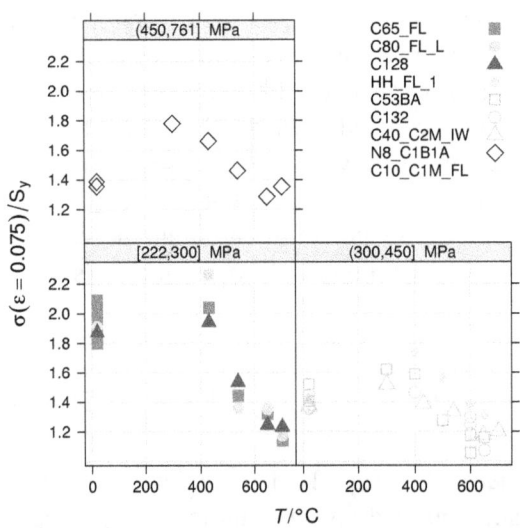

(b) in different room-temperature yield strength ranges

Figure 4: Ratio of the stress evaluated at $\epsilon = 0.075$ to the yield strength (a) as a function room-temperature yield strength, S_y^0. Sub-plots show four temperature ranges. (b) as a function of temperature. Sub-plots show different room-temperature yield strength ranges. The notation in the strips of the panels denotes a range of conditions, e.g. "(20,300] °C" denotes the range $(20 \leq T < 300)$ °C.

16

above about 500 °C increasing the strain rate by ten times can increase the measured strength 25 %; see for example Figure 36. In a constant-load test, this behavior manifests itself as creep, see for example Williams-Leir [38] for creep models for structural steels. During creep at constant load, continuous deformation occurs. The steel no longer requires an increase in stress to produce an increase in strain as it does at room temperature. Some literature reports have recognized that this behavior is important [9, 39, 40], but most [9, 39] have not incorporated it into models in a transparent manner.

The strain rates in standard high-temperature tensile tests [11, 36, 37] produce about 5 % strain in ten minutes. Therefore, the strengths measured in those tests represent an upper limit to the load-carrying capacity at that temperature. It is important, then, to be able to account for the time- or rate-dependent strength of steel. The increase in stress, σ, relative to a reference stress, σ_0 that results from an increase in strain rate at constant temperature can be expressed through another power-law relation:

$$\frac{\sigma}{\sigma_0} = f(\dot{\epsilon}) = \left(\frac{\dot{\epsilon}}{\dot{\epsilon_0}}\right)^m \tag{8}$$

The exponent, m, is termed the strain-rate sensitivity. The normalizing parameter, $\dot{\epsilon_0}$, ensures that the multiplier is unity at the strain rate at which the stress-strain behavior was measured. For the data of this report $\dot{\epsilon_0} = 8.33 \times 10^{-5}$ 1/s, the rate used in ASTM E8 [10].

The strain-rate sensitivities m, for the steels of this report, which were determined from the tensile tests, see Appendix D.4.1 and Table 6 for details, increase strongly with increasing temperature, Figure 5.

No one, simple, materials-science-based equation can describe behavior of the strain rate sensitivity for all temperatures and strain rates, because the mechanism responsible for deformation changes as temperature increases. Chapter 8 of Frost and Ashby's book [41] describes the deformation-mechanism-map approach for understanding the behavior of ferrous alloys. It breaks the temperature and strain-rate space into regimes where different deformation mechanisms dominate, and presents the governing equations for each regime. The deformation-mechanism-map approach, which is based on fundamental mechanisms, is too complicated for a simple representation. However, the general behavior of the mechanisms should be retained even in a phenomenological approach, so that the material response behaves well outside the range of the data. One important limitation is that strain-rate sensitivity for power-law creep of ferrous alloys has a maximum [42]: $m_{plc} \leq 0.2$, so any expression for the temperature dependence of the strain rate sensitivity should never exceed that maximum. A convenient, phenomenological expression that meets that requirement is

$$m(T) = m_0 + m_3 \left[1 - \exp\left(-\left(\frac{T}{m_2}\right)^{m_1} \right) \right] \tag{9}$$

Widely used, materials-science-based models for temperature and strain rate dependence of the stress-strain behavior, such as the Johnson-Cook [43] and Zerilli-Armstrong [44] models, employ temperature-independent strain-hardening exponents, n, and strain rate sensitivities, m. Of course, the data in Table 6 and the model, Eq. (3), could be used to generate the parameters of these models as well.

Figure 5 also compares the strain rate sensitivity, m, data of the steels of this study to literature data for structural steels, which are summarized in Appendix E.1. Although the strain rate sensitivities of the steels of this study were evaluated near an engineering strain, $e \approx 0.02$, the literature data in Figure 5 were calculated from the yield strength, since that parameter is most commonly reported. Figure 5 demonstrates that the strain rate sensitivities of the steels of this study are similar to other low-alloy structural steels, however.

To compute the most appropriate values for parameters in the the strain rate sensitivity relation, Eq. (9), the data of this report, Figure 27 were combined with existing literature data. The entire data set was fit using Eq. (9). Appendix E.5 details the procedure, and Table 3 summarizes the values. The solid line in Figure 5 is the Eq. (9) evaluated with the best-fit parameters.

4.2.5 Final form of the stress-strain behavior

The expressions for the retained strength R, the elastic modulus, E, strain-rate sensitivity, m, and the strain-hardening behavior can be combined into the final form of the stress-strain model.

$$\sigma = \begin{cases} E\epsilon & \text{for } \epsilon < S_y/E \\ \left(RS_y^0 + (k_3 - k_4 S_y^0) \exp\left(-(\frac{T}{k_2})^{k_1} \right) (\epsilon - RS_y^0/E)^n \right) \left(\frac{\dot{\epsilon}}{\dot{\epsilon_0}} \right)^m & \text{for } \epsilon \geq S_y/E \end{cases} \tag{10}$$

where

$$E = E_0 + e_1 T + e_2 T^2 + e_3 T^3 \qquad \text{Eq. (6)}$$

$$R = \exp\left(-\frac{1}{2}\left(\frac{T^*}{r_3}\right)^{r_1} - \frac{1}{2}\left(\frac{T^*}{r_4}\right)^{r_2} \right) \qquad \text{Eq. (1)}$$

$$m = m_0 + m_3 \left[1 - \exp\left(-\left(\frac{T}{m_2}\right)^{m_1} \right) \right] \qquad \text{Eq. (9)}$$

18

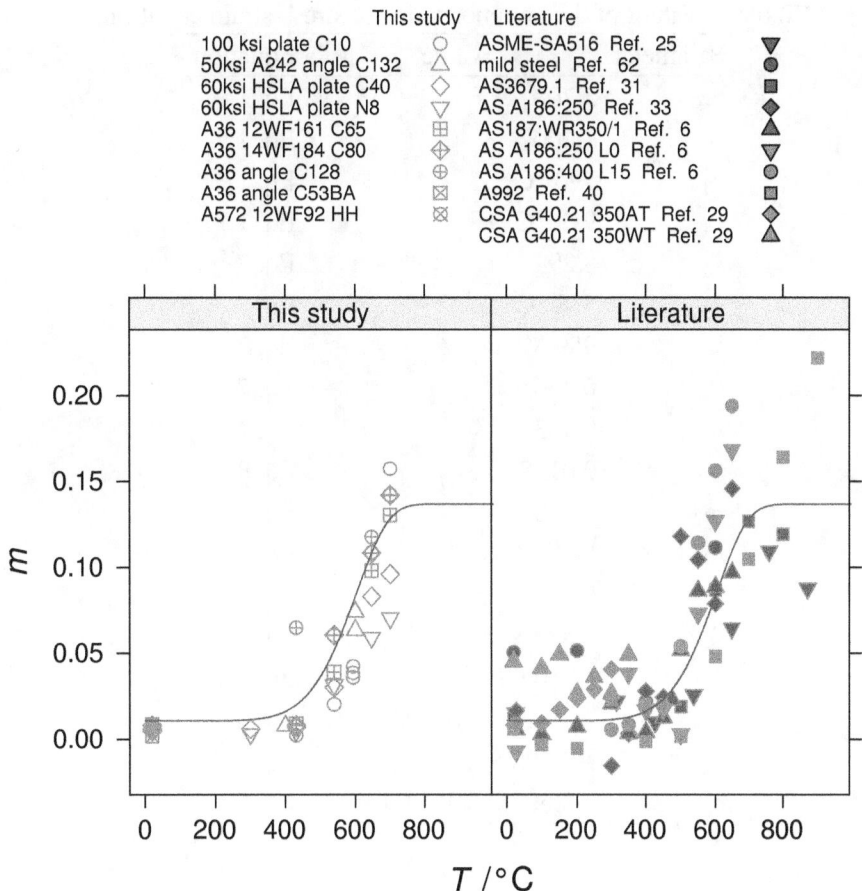

Figure 5: Strain rate sensitives, m, of WTC steels compared to literature data for structural steels.

For strain rates other than the reference strain rate, $\dot{\epsilon}_0$, the high-temperature yield strength, RS_y^0 should also be scaled by the strain rate sensitivity.

The five parameters, k_i and n, in Eq. (7) were estimated by a non-linear least-squares regression of the stress-strain data for 43 individual stress-strain curves, excluding the $F_y = 100$ ksi steels. Appendix D.5 summarizes details and the results of the fit. Table 3 summarizes the values. Figure 6 plots the relative deviation,

$$D_r = \frac{\hat{\sigma} - \sigma}{\sigma} \qquad (11)$$

between the plastic true stress predicted by Eq. (7), $\hat{\sigma}$, and the measured plastic

Table 3: Values of the parameters in the stress-strain equation.

Parameter	Value	Equation
r_1	5.708	Eq. 1
r_2	1.000	Eq. 1
r_3	590 °C	Eq. 1
r_4	919 °C	Eq. 1
k_1	8.294	Eq. 7
k_2	538 °C	Eq. 7
k_3	959 MPa	Eq. 7
k_4	0.766	Eq. 7
n	0.483	Eq. 7
m_0	0.0108	Eq. 9
m_1	7.308	Eq. 9
m_2	613 °C	Eq. 9
m_3	0.126	Eq. 9
$\dot{\epsilon}_0$	8.333×10^{-5} 1/s	Eq. 9
E_0	206.0 GPa	Eq. 6
e_1	-4.326×10^{-2} GPa/°C	Eq. 6
e_2	-3.502×10^{-5} GPa/°C^2	Eq. 6
e_3	-6.592×10^{-8} GPa/°C^3	Eq. 6

true stress, σ. The agreement is usually within 20 % and is frequently better. Appendix D.2 plots other comparisons.

4.3   Comparison with Eurocode recommended values

The Eurocode 3 standard for structural fire design [1] defines, for design purposes, the stress-strain behavior of structural steel at elevated temperatures. Its model uses four temperature-dependent parameters to describe the engineering stress-strain ($S - e$) curve:

- the Young's modulus ($\bar{E}_{a,\theta}$),

- the proportional limit ($f_{\mathrm{ap},\theta}$),

- the stress at $e = 0.02$ ($f_{\mathrm{amax},\theta}$), and

- the tensile strength $f_{\mathrm{au},\theta}$.

An ellipse connects the stress-strain point at the proportional limit to the maximum stress, $f_{\mathrm{amax},\theta}$. The temperature dependence of each parameter is expressed in a form normalized to a room-temperature value. The form is not a smooth function; instead it specifies fixed values at 100 °C increments. The stress at $e = 0.02$, $f_{\mathrm{amax},\theta}$, is normalized to the room-temperature $e = 0.002$ offset yield strength, $S_y(0.2\ \%\ \text{offset})$, from the mill-test report. This normalizing value is neither the room-temperature proportional limit, $f_{\mathrm{ap},\theta}$, nor the room-temperature stress at $e = 0.02$, $f_{\mathrm{amax},\theta} = S_y(2\ \%\ \text{extension})$. Appendix F explains the Eurocode 3 stress-strain model in greater detail.

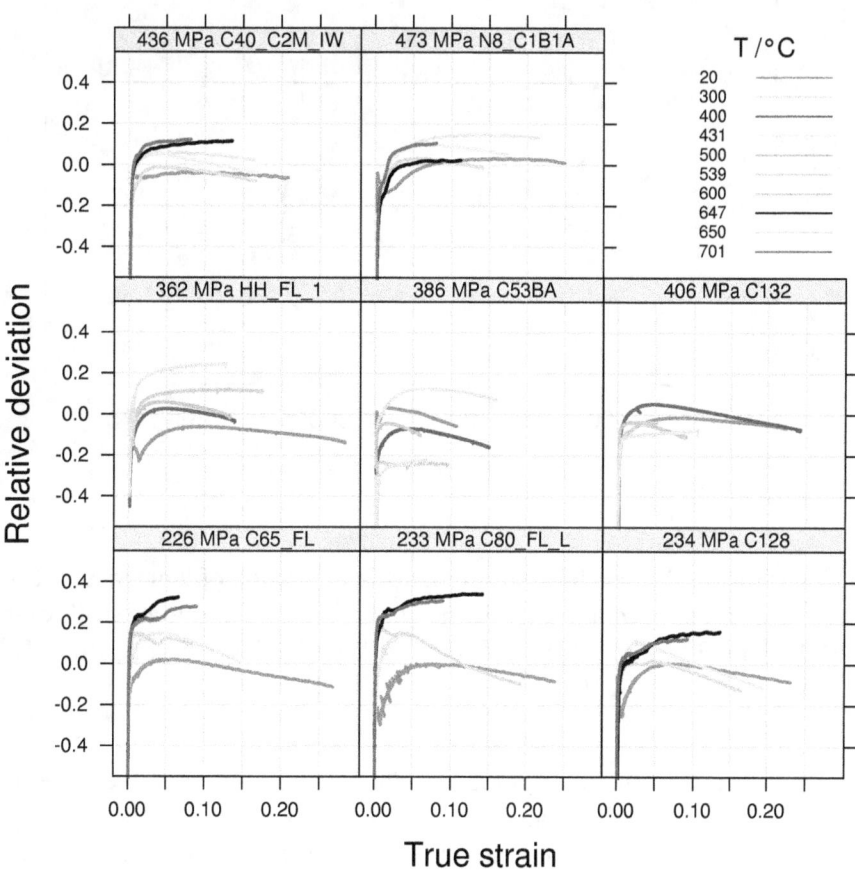

Figure 6: Relative deviation, D_r, between true stress predicted by Eq. (7) and the measured stress.

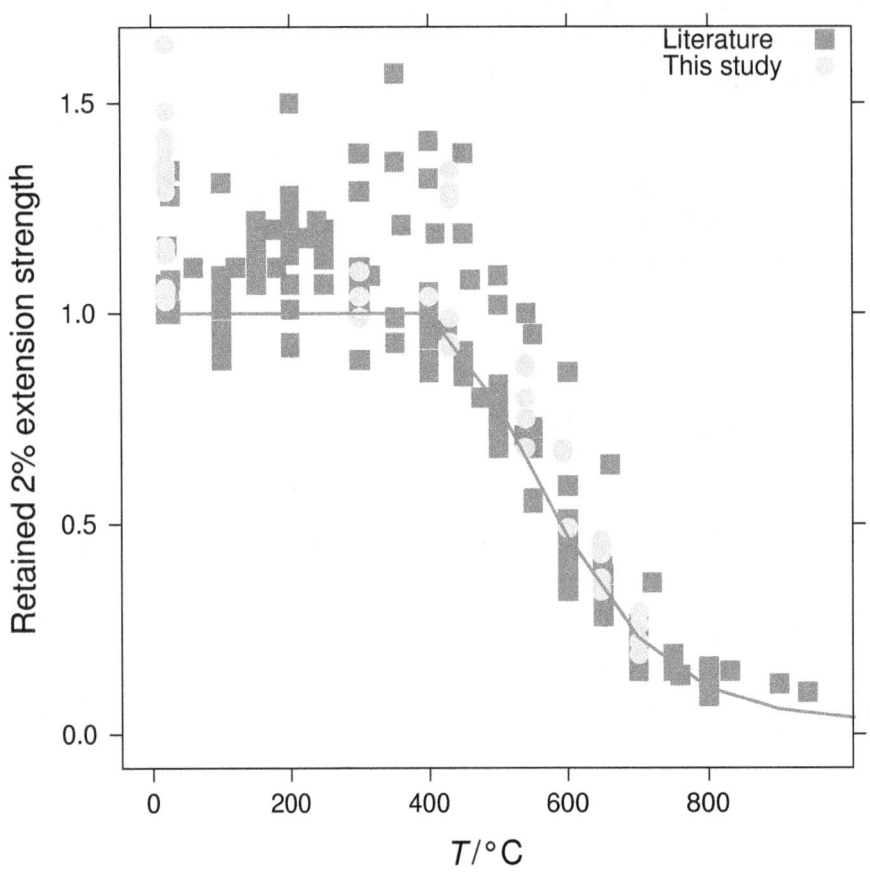

Figure 7: Normalized $e = 0.02$ yield strengths as a function of temperature. Solid line is the Eurocode 3 recommended value. Data are from appropriate steels of this study (C10_C1M_FL, C128, C132, C40_C2M_IW, C53BA, C65_FL, C80_FL_L, N8_C1B1A) and literature [6, 12, 18, 23, 25, 28, 30–33] .

Figure 7 compares the rate-appropriate data of this study and literature data [6, 12, 18, 23, 25, 28, 30–33] to the Eurocode 3 recommended value for $f_{\mathrm{amax},\theta}$. All the data in Figure 7 are from tests in the strain-rate range $3.30 \times 10^{-5} < \dot{\epsilon} < 1.35 \times 10^{-4}$. All the data for the steels of this study lie above the Eurocode 3 line at room temperature because they all strain-harden: $S_y(2\,\%\ \mathrm{extension}) > S_y(0.2\,\%\ \mathrm{offset})$. Most lie above the recommended value for temperatures up to 650 °C. Appendix E.1 summarizes the literature steels in Figure 7.

It is also possible to compare the prediction of the Eurocode 3 stress-strain model to the stress-strain data. Figure 8 plots the relative deviation, D_r between the data and the Eurocode 3 stress-strain prediction. The figure is is directly comparable to Fig. 6. Note that the Eurocode prediction for the five lowest strength steels is consistently lower than the actual data. Appendix F.3 compares the pre-

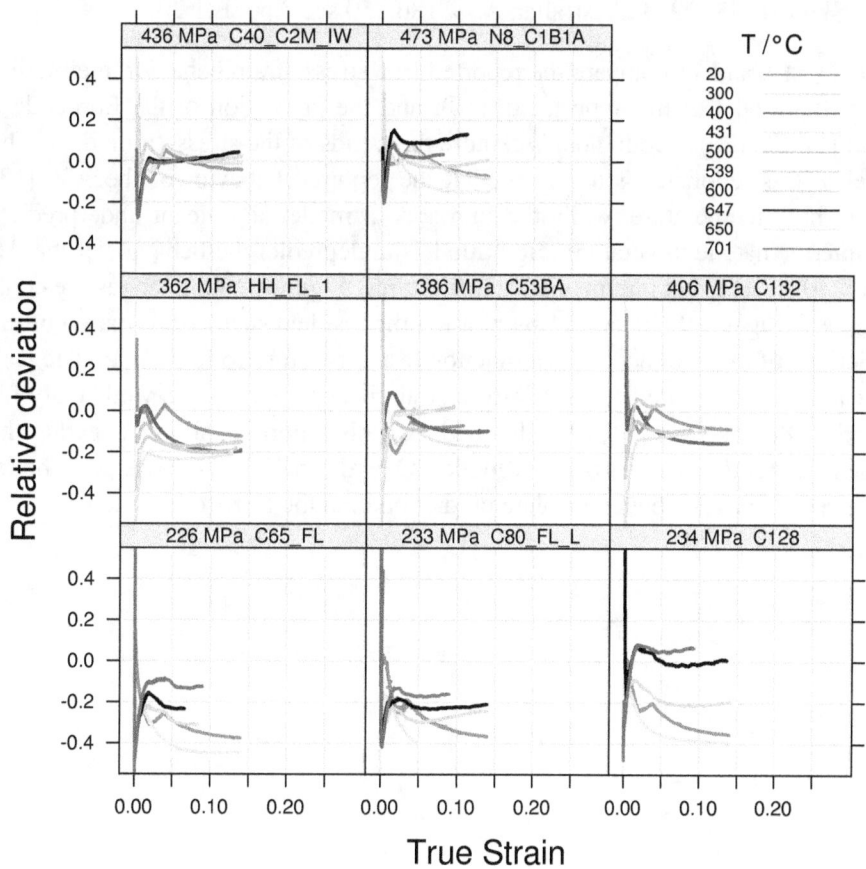

Figure 8: Relative deviation, D_r between true stress predicted by the Eurocode 3 stress-strain model, Eq. (20), and the observed behavior.

dictions of the model of this report, Eq. (10) and the Eurocode 3 model to the actual stress-strain data for each steel individually.

Another method for understanding the utility of the model is to compare its predictions on steels not analyzed as part of this report. Relatively few [13, 33,

40, 45] open-literature sources exist that report full stress-strain data for structural steels in a form suitable for graphical comparison. Of these, three literature reports that document stress-strain behavior were chosen:

- Hu [40] 2009 (ASTM A992) see App. E.1.10,

- Harmathy [13] 1970 (ASTM A36), see App. E.1.7, and

- Skinner [33] 1973 (Australian AS A186:250 see App. E.1.20).

Figures 9, 10, and 11 compare the reported true stress-strain behavior, the prediction of the model of this report, Eq. (10), and the prediction of the Eurocode 3 model [1], Eq. (20). Both models capture the trends of the stress-strain data. The model of this report tends to overpredict the reported behavior of the A36 [13] steel at high temperature, while the Eurocode 3 model significant underpredicts it at intermediate tepmperature. Both models undepredict the behavior of the AS A186:250 steel [33] at intermediate temperatures. Figure 12 summarizes the comparisons of Figures 9, 10, and 11 by plotting the residual standard deviation of the predictions of the two models as a function of temperature for $\epsilon > 0.02$. The lack of clear trend in the predictions of stress-strain is apparent here as well. Only for the AS A186:250 steel [33] does the model of this report consistently predict the behavior better than the Eurocode 3 model. Overall, in 12 of 24 cases , the model of this report makes a better absolute prediction than the Eurocode 3 model.

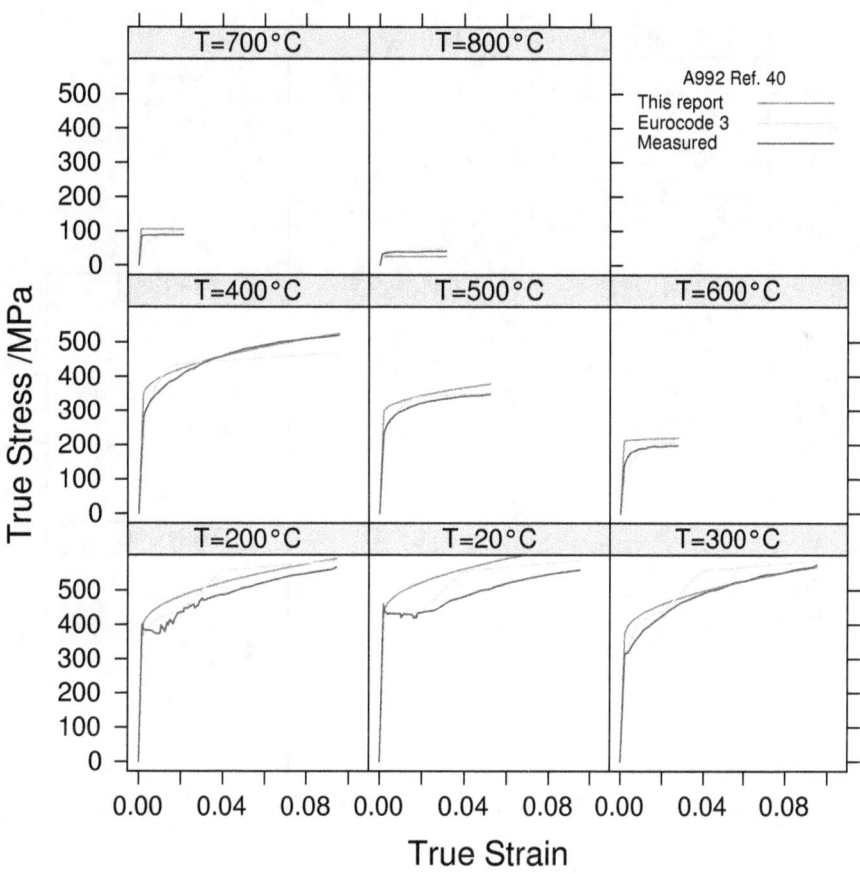

Figure 9: Comparison of reported stress-strain data, prediction of Eq. (10) and the Eurocode 3 [1] stress-strain model for 2000-era A992 steel [40] with $F_y = 345$ MPa (50 ksi). Individual panels show test temperatures.

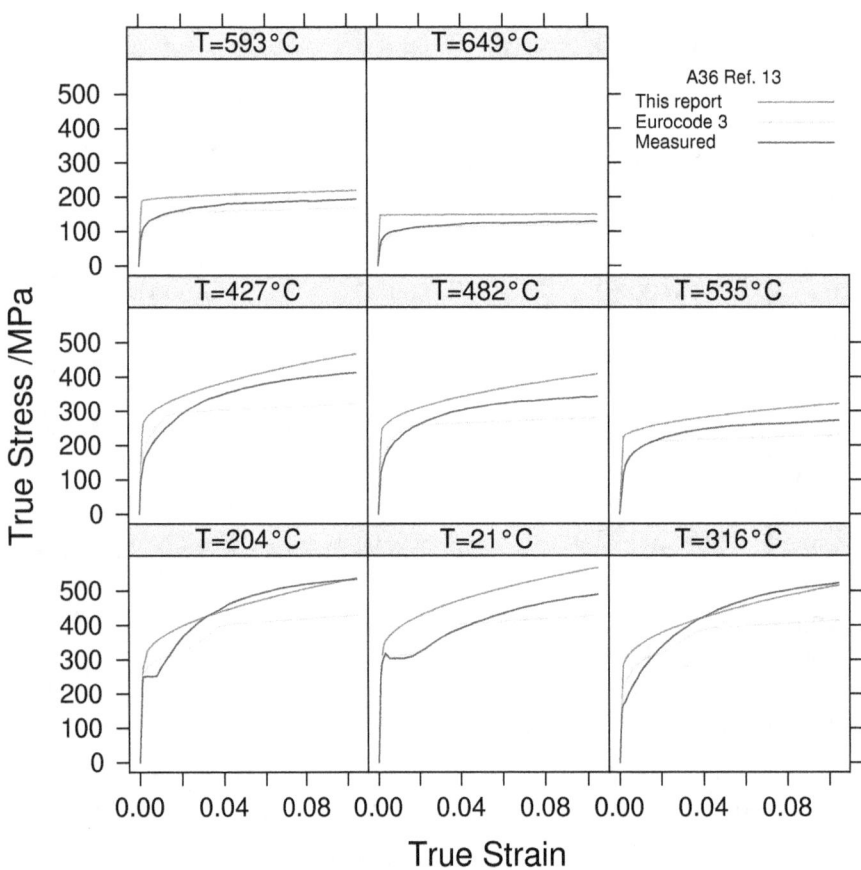

Figure 10: Comparison of reported stress-strain data, prediction of Eq. (10) and the Eurocode 3 [1] stress-strain model for 1970s-era A36 steel [13] with $F_y = 248$ MPa (36 ksi). Individual panels show test temperatures.

26

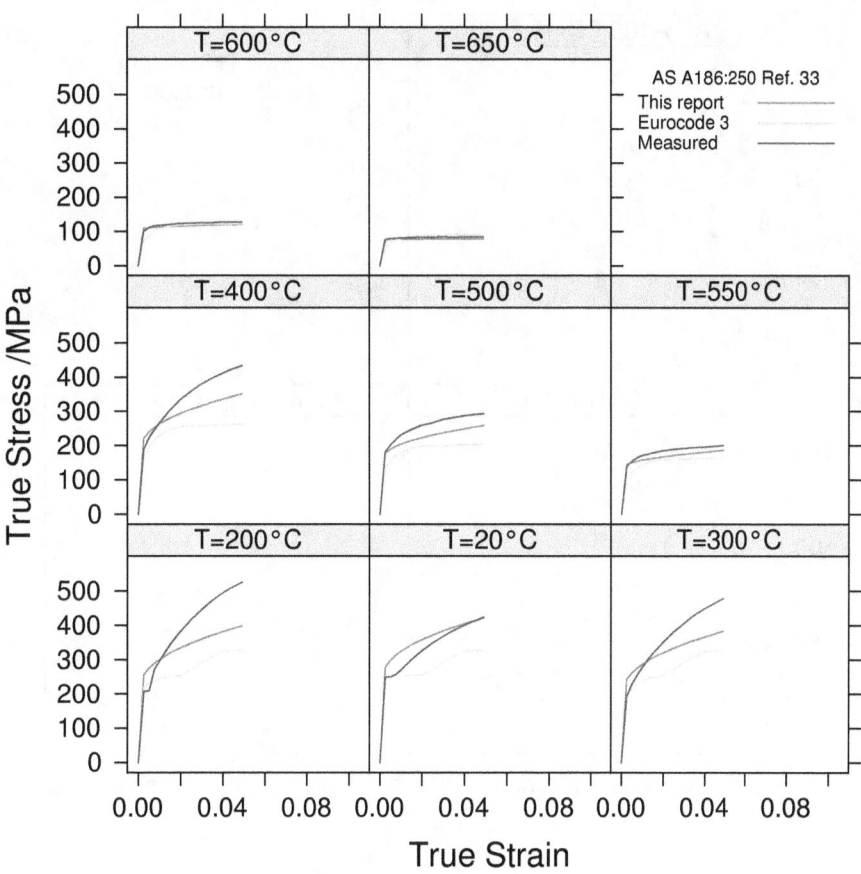

Figure 11: Comparison of reported stress-strain data, prediction of Eq. (10) and the Eurocode 3 [1] stress-strain model for Australian 1970s-era AS A186:250 steel with $F_y = 250$ MPa. Individual panels show test temperatures.

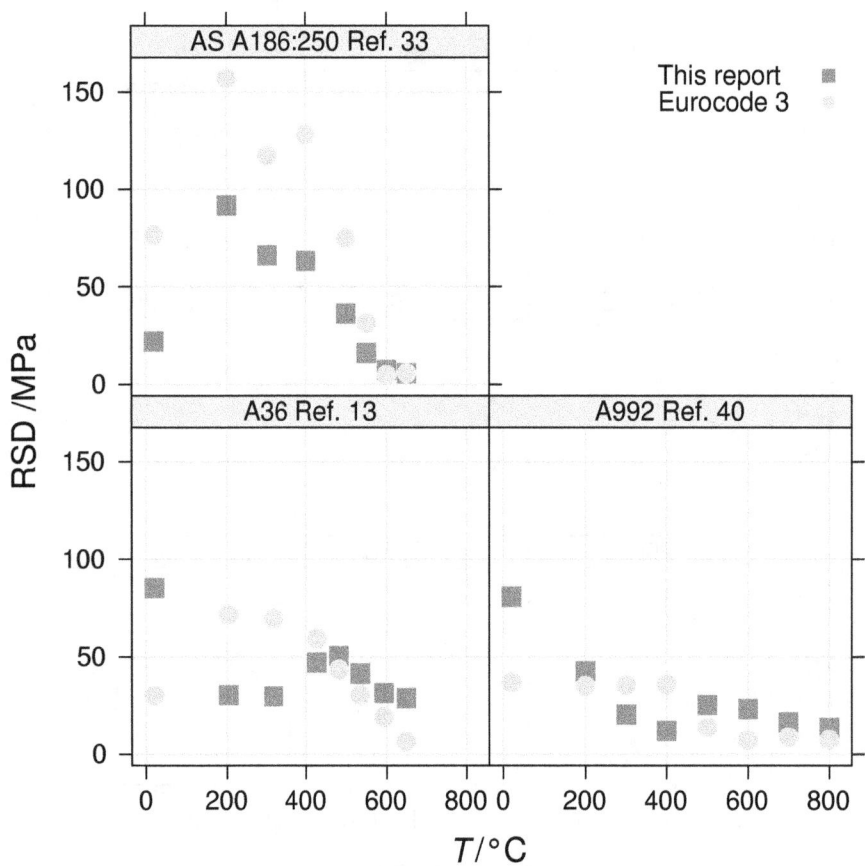

Figure 12: Comparison of absolute average deviation of the reported stress-strain data for $\epsilon \geq 0.02$ from the prediction of Eq. (10) and the Eurocode 3 [1] stress-strain model

28

5 Summary

This report documents a model to represent the true stress-strain, $\sigma - \epsilon$, behavior of structural steel. It is based on combination of data from the NIST World Trade Center collapse investigation and many other evaluated literature sources. Unlike other models for stress-strain behavior of structural steel, such as the Eurocode 3 formulation [1], the model explicitly describes the time-dependent nature of the strength of steel at high temperature. For untested steels, it predicts the stress-strain behavior using only the measured room-temperature yield strength, S_y^0. The relative deviation between the model of this report and the actual data for the steels is generally less than 25 %, and is always less than 50 %. On subset of eight steels, the model predicts the stress-strain behavior slightly better than the equally complicated Eurocode 3 model. For three literature structural steels, not analyzed as part of the model, the model of this report and the Eurocode 3 model predict stress-strain behavior with similar quality.

Acknowledgements and contributions

WEL performed the tensile tests for C53BA, C132, and HH, analyzed the data for discussion section, and wrote the manuscript. JDM performed the tensile tests of C65, C80, C128, C40, N8, and C10. SWB characterized the microstructures and managed the chemical analysis, Appendix 3.1.

A   Symbols and Definitions

Table 4 below summarizes the symbols used in this report.

Table 4: Symbols used in this manuscript.

Symbol	Definition
ϵ	true strain; natural log of the current length, l over original length, l_0: $\epsilon = \log_e(\frac{l}{l_0})$
e	engineering strain: $e = \dfrac{l - l_0}{l_0}$
$\dot{\epsilon}$	true strain rate
σ	true stress: force, P, over current area, A: $\sigma = \dfrac{P}{A}$
S	engineering stress: force, P over original area, A_0: $S = \dfrac{P}{A_0}$
S_y	measured yield strength as defined in ASTM E6 [46]. Sometimes denoted YS in other sources. Can be qualified with the type of yield strength. For example S_y (0.2 % offset) is the traditional offset yield strength in mill test reports.
F_y	in this report, F_y denotes the specified, as opposed to measured, yield strength, for example, a "$F_y = 36$ ksi steel." This usage is consistent with the AISC [47] Steel Design Manual.
S_y^0	measured yield strength at room temperature.
S_u	tensile strength as defined in ASTM E6 [46]. Often called "ultimate tensile strength," and defined as the maximum force over the original specimen area: $P_{\max}/A_0$. Often abbreviated in the literature as TS.
El_t	Total elongation to failure as defined in ASTM E6 [46]
RA	Reduction of area at failure, a measure of the ductility, as defined in ASTM E6 [46]
K	prefactor that defines the strength of the steel in the stress-strain curve, defined in Eq. (13), and often called the strength coefficient.
n	strain-hardening exponent defined in Eq. (13)
m	strain rate sensitivity defined in Eq. (8) and (9)
R	retained strength, usually yield strength, S_y, normalized to its room-temperature value, S_y^0.
D_r	relative deviation of stress from model, Eq. (11)

continued on next page

Table 4: Symbols used in this manuscript

Continued from the previous page

Symbol	Definition
D_a	absolute deviation of stress from model, Eq. (16)
$\bar{E}_{a,\theta}$	Young's modulus in the Eurocode 3 standard, Eq. (20)
$f_{ap,\theta}$	the proportional limit in the Eurocode 3 standard, Eq. (20)
$f_{amax,\theta}$	stress at $\epsilon = 0.02$ in the Eurocode 3 standard, Eq. (20)
$f_{au,\theta}$	stress at $\epsilon = 0.04$ in the Eurocode 3 standard for capturing strain hardening, Eq. (20)
$k_{E,\theta}$	non-dimensional reduction factor for Young's modulus in the Eurocode 3 standard, Eq. (25)
$k_{p,\theta}$	non-dimensional reduction factor for the proportional limit in the Eurocode 3 standard, Eq. (26)
$k_{max,\theta}$	non-dimensional reduction factor for the stress at $\epsilon = 0.02$ in the Eurocode 3 standard, Eq. (27)
$k_{u,\theta}$	non-dimensional reduction factor for the stress at $\epsilon = 0.04$ in the Eurocode 3 standard for capturing strain hardening, Eq. (28)
RSD	residual standard deviation of a fit, defined in Ref. 48

$$RSD = \left(\frac{1}{\nu} \sum_{i=1}^{i=n} (y - \hat{y})^2 \right)^{0.5} \tag{12}$$

where $\hat{y}$ is the predicted value from the model, and ν is the number of degrees of freedom of the fit to n data points. Smaller values represent better fits.

ν	Number of degrees of freedom in a regression

B Experimental Methods

B.1 Materials and Microstructural Characterization

A commercial testing laboratory characterized the chemistry of the nine specimens using optical emission spectroscopy. Tests followed ASTM E415 [49] for all elements except C, which was analyzed in accord with ASTM E1019 [50].

Standard metallographic procedures were used to prepare the samples for optical microscopy. Specimens were ground with SiO_2 papers to 800 grit and then polished with diamond pastes of 6 μm, 1 μm, and 0.25 μm. The final, hand-polishing step used a colloidal mixture of 0.05 μm SiO_2 and Al_2O_3 particles. The microstructures were revealed using a combination of 4 percent picral (4 g picric acid, 96 mL ethyl alcohol) followed by 2 percent nital (2 mL concentrated nitric acid, 98 mL ethyl alcohol).

The volume fraction of pearlite, f_v^p, in the hot-rolled steels was determined using an area-detected measurement by automatic image analysis techniques using seven measurements along the plate centerline at 200x or 500x magnification.

B.2 Tensile testing

High-temperature tensile tests employed two different machines. One was an electromechanical testing machine with a contact extensometer that had a 12.5 mm gage length. The furnace was a split, $MoSi_2$-element design. The specimen for this machine, designated Flat C3 in Figure 13, conformed to Fig. 1 of ASTM E 21. [11] The uniform cross-section was 32 mm long and 3.175 mm by 6.0 mm wide. They were loaded using pin-aligned, superalloy, wedge grips. The specimen temperature was monitored using a K-type thermocouple mounted within 1 mm of the specimen surface. The specimens were loaded as soon as possible after the furnace temperature stabilized at the test temperature, which was generally within 20 minutes. During the temperature stabilization step, the specimen temperature was always within 5 °C of the desired temperature. The actuator displacement rate in these tests was 0.0325 mm/s, which produced a strain rate after yielding of approximately 0.001 s^{-1}, measured using the extensometer. Prior to yield, the actuator displacement rate produced a specimen stressing rate of (4 ± 2) MPa/s. These rates meet the requirements of ASTM E 8 [10] and E 21. [11]

Most other tests used a second electromechanical testing machine with a high-temperature extensometer with a 25.4 mm gage length. The furnace was a split design in which quartz lamps heated the specimen. This test frame employed two specimen types. The round specimens, designated Rd 1 in Fig. 13, had a 38 mm uniform long cross section 6.35 mm in diameter. The flat specimens, designated Flat C1 in Fig. 13, had a uniform cross-section 28 mm long and 6.35 mm wide. The thickness was that of the original plate, typically also 6.35 mm. These tests followed the loading protocol of ASTM E 21. [11] The initial actuator displacement rate was 0.00167 mm/s. At about $\epsilon = 0.02$, the operator increased the actuator displacement rate to 0.0167 mm/s, as required by E 21. The extension rate jump produces a discernible step in the flow stress, $\Delta\sigma$, for elevated temperature tests. The presence of this stress jump, $\Delta\sigma$, made it possible to evaluate the strain rate sensitivity of the stress for each temperature. Appendix D.4.1 summarizes the method used to estimate the strain rate sensitivity from the stress jump. Tests on three of the nine steels employed only a single extension rate, so it was impossible to evaluate their strain rate sensitivity.

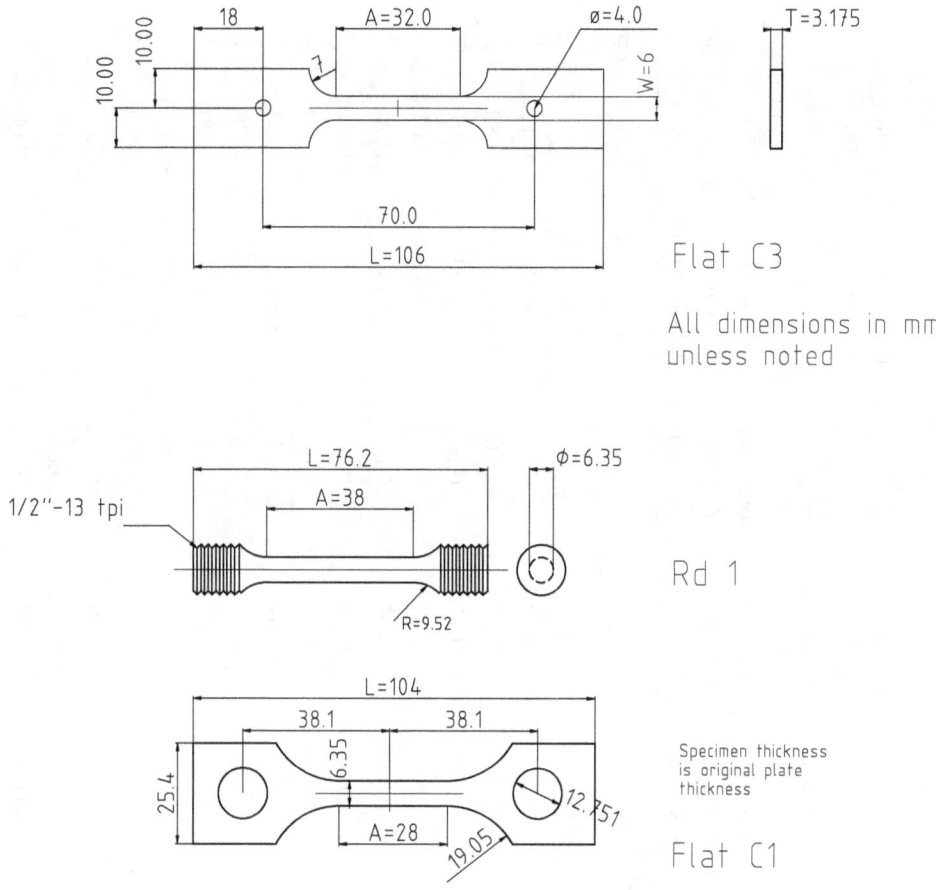

Figure 13: Tensile test specimens used in high-temperature tests.

C   Microstructure and chemistry of the steels

Table 5 summarizes the origin and chemistry of the nine steels. All of the recovered steels are low-carbon alloys with carbon contents below 0.25 percent by mass, strengthened by common elemental alloying practices. Carbon and Manganese are the main strengthening elements in the wide flange core columns, floor trusses, and floor truss seat. The floor truss angles and higher strength core column, HH, also have minor additions of V for carbide formation. The two lower-strength perimeter column plates, C40 and N8, originated from an inner web and a flange respectively. Specimen C40 contains V, while flange plate N8 contains Nb. Their significantly different chemistries, e.g., Si, Cu, V, Nb, suggest that different mills supplied the two plates. Contemporaneous construction documents indicate that domestic mills

Table 5: Chemistry and description of the steels

Abbreviation	C65	C80	C128	HH	C53BA	C132	C40	N8	C10
Specimen	C65-fl-1	C80-fl-1	C128-T1	HH-fl-1	C53-ba3	C132-ta-3	C40-c2m-iw-1	N8-c1b-f-1	C10-c1m-fl-1
F_y (ksi)	36	36	36	42	36	50	60	60	100
Description of shape	12WF161 flange	14WF184 flange	channel	12WF92 flange	(76×50) mm angle	(50×28) mm angle	plate	plate	plate
Source†	904:86-89	603:92-95	unknown	605:89-101	unknown	unknown	136:98-101	142:97-100	451:85-88
t	37.7 mm	35 mm	ND	22 mm	9.4 mm	6.35 mm	6.35 mm	7.9 mm	6.35 mm
f_v^{p*}	0.282 (0.019)	0.366 (0.027)	0.244 (0.022)	0.247 (0.018)	0.266 (0.009)	0.288 (0.007)	0.264 (0.032)	0.256 (0.019)	NA
Chemistry									
Element†	x_i %	x_i %	x_i %	x_i %	x_i %	x_i %	x_i %	x_i %	x_i %
C	0.23	0.23	0.19	0.17	0.17	0.19	0.15	0.16	0.17
Mn	0.74	0.90	0.83	1.08	0.60	0.82	1.10	1.50	0.89
P	0.006	0.008	0.010	<0.005	0.013	0.010	0.006	0.023	0.015
S	0.015	0.008	0.005	0.014	0.040	0.029	0.023	0.016	0.012
Si	0.02	0.03	0.04	0.03	0.02	0.07	0.03	0.44	0.24
Ni	0.02	0.01	0.01	0.02	0.09	0.08	0.08	0.01	0.01
Cr	0.02	0.02	0.02	0.02	0.08	0.10	0.06	0.02	0.90
Mo	0.01	0.01	0.02	<0.01	0.01	<0.01	<0.01	<0.01	0.43
Cu	0.05	0.05	0.03	0.24	0.32	0.32	0.23	0.05	0.28
V	<0.005	<0.005	0.002	0.065	0.039	0.038	0.052	0.005	0.007
Nb	<0.005	<0.005	ND	<0.005	<0.005	<0.005	<0.005	0.064	<0.005
Ti	<0.005	<0.005	ND	<0.005	<0.005	<0.005	<0.005	0.007	<0.005
Zr	<0.005	<0.005	ND	<0.005	<0.005	<0.005	<0.005	<0.005	<0.01
Al	<0.005	<0.005	ND	<0.005	ND	0.032	<0.005	0.042	0.081
B	<0.0005	<0.0005	ND	<0.0005	<0.0005	<0.0005	<0.0005	<0.0005	<0.0005
N	0.0041	0.0040		0.0099	0.0090	0.0080	0.0057	0.0033	0.0040

Notes: ND = not determined; NA = not applicable. ‡Chemical composition, x_i expressed in mass fraction. t is the original thickness

* f_v^p: Volume fraction of pearlite. Uncertainty listed in parentheses is the standard deviation of 7 measurements.

† Source : all identified sections from WTC 1 ; "904" identifies the column line; "86-89" identifies the range of floors.

supplied the inner web plates and Yawata Iron and Steel (now Nippon Steel) supplied all the flange plates [51]. The higher strength flange plate, C10, contains elevated levels of Cr and Mo for increased strength and hardenability.

Figures 14 g-i show the microstructures from the three hot-rolled, wide-flange columns. The ferrite (white) and pearlite (gray-black) are uniformly distributed across the sections, although lower magnification images revealed some localized banding of the two phases. The ferrite morphology is a combination of polygonal and irregular. Some Widmanstätten morphology is present, though not shown, in the two $F_y = 36$ ksi* columns, C65 and C80. The ferrite grain size in the $F_y = 42$ ksi column is finer than that in the two lower-strength core columns, which appears to have a larger effect on the strength of the material than the volume fraction of pearlite, f_v^p, see Table 5.

The microstructures of the truss seat and the two floor truss angles are shown in Figures 14 d–f. All three steels are hot-rolled. The ferrite and pearlite are uniformly distributed across the sections. The ferrite morphology is primarily irregular or polygonal. The pearlite volume fractions, f_v^p, are similar, Table 5. Although the same mill produced the two angles, the ferrite grain size and volume fraction of Widmanstätten morphology are larger in the lower strength angle, C53, than in the higher strength angle, C132.

Figures 14 a–c show the microstructures from the perimeter column plates. Both $F_y = 60$ ksi plates, inner web C40 and flange N8, plates are hot-rolled steels with a ferrite-pearlite microstructure. Although different mills rolled the plates, their microstructures are similar. Both the ferrite and pearlite are moderately banded, particularly near the centerline of the plate. This gradient in banding usually arises from compositional gradients present in the original ingot. The ferrite morphology is a mixture of polygonal, irregular, and Widmanstätten. The ferrite grain sizes of the two plates are similar, and volume fractions of pearlite, f_v^p, are statistically equivalent. Two pearlite morphologies exist, depending upon the local banding characteristics. In unbanded areas the pearlite colonies contain dense, but distinguishable, lamellar plates. In the more heavily banded regions, the pearlite appears mottled. The lamellar spacing of this pearlite could not be resolved in the optical microscope. Some areas may be degenerate pearlite, but it was not possible to be certain using only the optical microscope. Occasionally, the pearlite morphology is granular. Bainite may also be present in regions where the ferrite is associated with small cementite particles, but again, optical microscopy does not permit definitive identification.

---

*This manuscript identifies several steels by their strength grade, expressed in English, rather than SI, units. The term "36 ksi" describes the strength class of the steel, rather than an exact measure of its strength. This manuscript reports measured strengths in units of MPa as required by the NIST Administrative Manual, Subchapter 4.09 Appendix D.

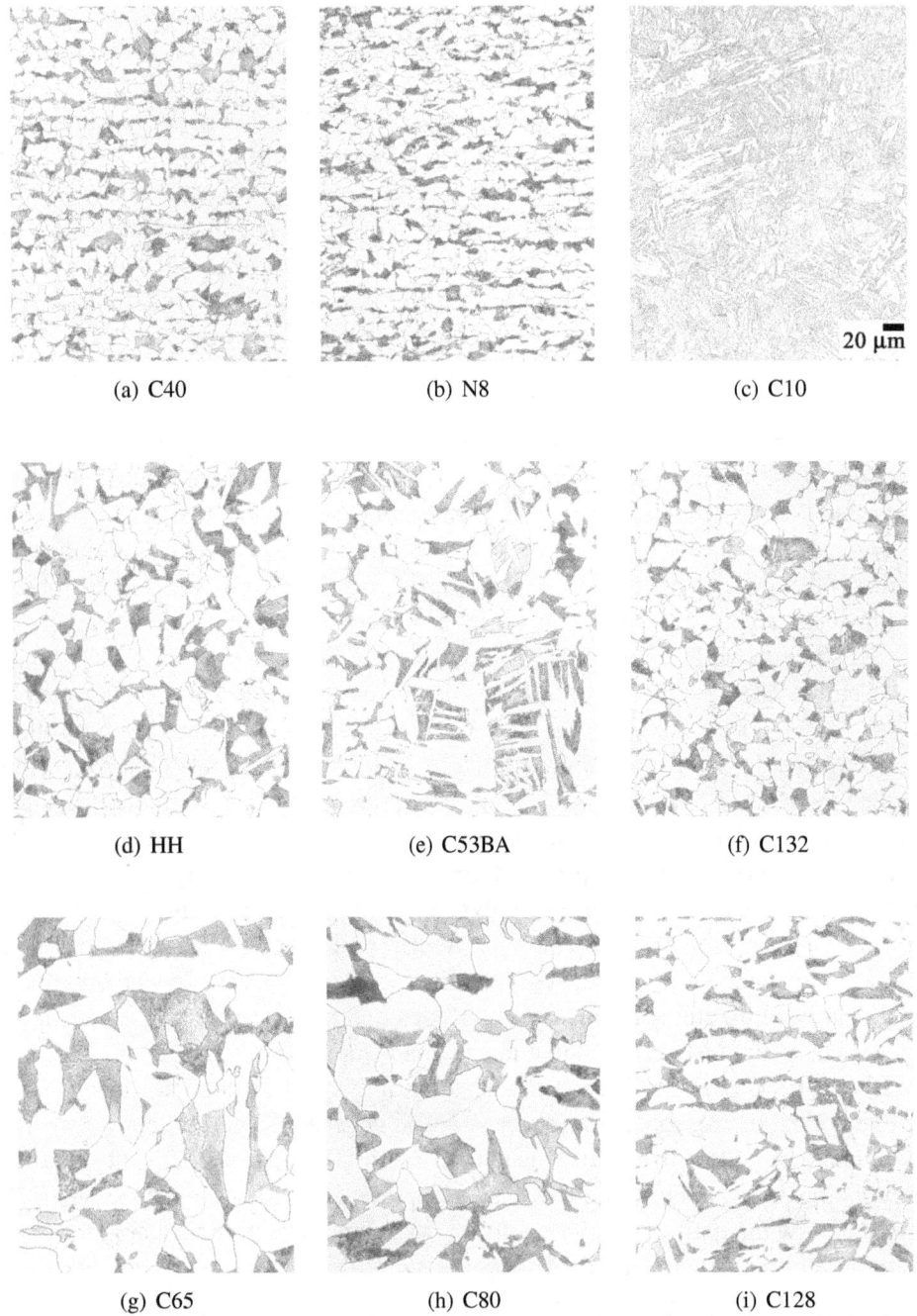

(a) C40   (b) N8   (c) C10

20 µm

(d) HH   (e) C53BA   (f) C132

(g) C65   (h) C80   (i) C128

Figure 14: Light optical micrographs of representative microstructures of the nine steels at the same magnification. Images were taken near the centerline of the plates on the transverse plane. The rolling direction is horizontal. Steels are ordered left-to-right and bottom-to-top by increasing measured yield strength. All steels except (c) C10 have ferrite-pearlite microstructures.

36

The $F_y = 100$ ksi flange plate C10, Fig. 14c, is a quenched-and-tempered steel. In the optical microscope, the microstructure appears to be a tempered martensite. Prior austenite grain boundaries and distinct remnants of ferrite lath boundaries are visible. Cementite carbides lie on the prior austenite grain boundaries and the lath boundaries.

D Supporting data, plots, and analysis

D.1 Tensile data

Table 6 summarizes the data used in this report. The footer of the table describes the symbols. The reported total elongations, El_t should be interpreted with some care. In general, they are measured over a gauge length $G = 25.4$ mm for high-temperature specimens, see Fig. 13. Room-temperature tests generally employed a specimen with $G = 50.8$ mm. Reductions of area, RA, for rectangular specimens were calculated by averaging the thicknesses, using the procedure in ASTM E8 [10]. Table 6 includes columns for the parameters of a power-law fit to the true stress-strain data:

$$\sigma = K\epsilon^n \tag{13}$$

The fit for K and n used the total, as opposed to plastic, true strain. The method for computing the strain-rate sensitivity, m, is summarized in Appendix D.4.1. The final column, labeled "lm" indicates whether the stress-strain curve was used in computing the parameters of the stress-strain model, Eq. (10).

Table 6: Data for steels of this study.

Specimen	T	$\dot{\epsilon}_0$	S_y^{002}	S_y^{02}	S_u	El_t	RA	K	n	m	lm
	°C	1/s	MPa	MPa	MPa	%	%	MPa			
C65	20	8.75×10^{-5}	225	320	461	43.0	62.7	815	0.234	0.0078	0
C65	20	8.75×10^{-5}	210	305	448	42.0	61.4	802	0.241	0.0069	0
C65	20	8.75×10^{-5}	215	313	483	43.0	62.2	817	0.239	0.0081	0
C65	20	8.75×10^{-5}	264	370	469	38.0	63.1	771	0.187	0.0074	0
C65	20	8.75×10^{-5}	240	334	455	42.0	61.8	777	0.212	0.0066	1
C65	20	8.75×10^{-5}	202	298	441	43.0	61.6	792	0.244	0.0087	0

NA: not available; data not calculated

$S_y^{002} = S_y(0.002 \text{ offset})$: 0.2 % offset yield strength

$S_y^{02} = S_y(0.02 \text{ elong})$: stress measured at 2 % total strain

S_u = tensile strength; El_t = total elongation; RA = reduction of area

lm: used in calculation of stress-strain model. 1=used, 0=not used

continued on next page

Table 6: Data for WTC steels of this study.

Continued from the previous page

Specimen	T	$\dot{\epsilon}_0$	S_y^{002}	S_y^{02}	S_u	El_t	RA	K	n	m	lm
	°C	1/s	MPa	MPa	MPa	%	%	MPa			
C65	20	2.19×10^{-3}	223	308	452	42.0	62.0	808	0.236	NA	0
C65	431	4.37×10^{-5}	194	302	392	42.0	71.8	685	0.212	0.0091	1
C65	539	4.37×10^{-5}	158	197	231	43.0	74.4	304	0.112	0.0391	1
C65	647	4.37×10^{-5}	85	98	128	91.0	96.5	141	0.087	0.0981	1
C65	701	4.37×10^{-5}	54	59	76	NA	96.6	68	0.040	0.1303	1
C80	20	8.75×10^{-5}	233	307	464	38.0	68.5	864	0.256	0.0047	1
C80	431	4.37×10^{-5}	190	300	407	37.0	73.1	881	0.276	0.0082	1
C80	539	4.37×10^{-5}	166	205	255	42.0	72.1	272	0.074	0.0604	1
C80	647	4.37×10^{-5}	86	105	135	61.0	97.6	146	0.088	0.1082	1
C80	701	4.37×10^{-5}	58	NA	88	83.0	98.0	78	0.050	0.1419	1
HH	20	1.71×10^{-4}	362	394	512	38.0	70.3	840	0.185	NA	1
HH	400	1.71×10^{-4}	270	374	449	NA	75.2	734	0.172	NA	1
HH	500	1.71×10^{-4}	239	315	352	NA	64.0	528	0.133	NA	1
HH	600	1.71×10^{-4}	158	200	208	NA	NA	268	0.078	NA	1
HH	650	1.71×10^{-4}	136	164	166	NA	84.0	214	0.072	NA	1
C128	20	4.37×10^{-5}	234	313	459	40.0	67.5	844	0.252	0.0088	1
C128	431	4.37×10^{-5}	190	298	401	37.0	70.6	608	0.192	0.0648	1
C128	539	4.37×10^{-5}	134	176	221	46.0	79.2	286	0.127	0.0603	1
C128	647	4.37×10^{-5}	72	80	108	68.0	97.1	112	0.084	0.1177	1
C128	701	4.37×10^{-5}	44	51	68	79.0	91.6	62	0.051	0.1421	1
C53BA	20	5.80×10^{-5}	413	498	560	26.0	58.1	805	0.126	0.0021	1
C53BA	20	1.57×10^{-3}	360	452	519	24.0	58.5	811	0.152	NA	0
C53BA	300	1.57×10^{-3}	370	468	579	NA	NA	947	0.176	NA	1
C53BA	400	1.02×10^{-3}	284	370	427	NA	54.3	658	0.146	NA	1
C53BA	500	1.02×10^{-3}	284	326	332	11.0	NA	445	0.080	NA	1
C53BA	600	1.02×10^{-3}	186	208	205	8.0	29.6	246	0.045	NA	0
C53BA	600	1.02×10^{-3}	183	186	183	16.0	28.1	207	0.026	NA	1

NA: not available; data not calculated

$S_y^{002} = S_y(0.002 \text{ offset})$: 0.2 % offset yield strength

$S_y^{02} = S_y(0.02 \text{ elong})$: stress measured at 2 % total strain

S_u = tensile strength; El_t = total elongation; RA = reduction of area

lm: used in calculation of stress-strain model. 1=used, 0=not used

continued on next page

38

Table 6: Data for WTC steels of this study.

Continued from the previous page

Specimen	T	$\dot{\epsilon}_0$	S_y^{002}	S_y^{02}	S_u	El_t	RA	K	n	m	lm
	°C	1/s	MPa	MPa	MPa	%	%	MPa			
C53BA	650	1.02×10^{-3}	126	140	138	20.0	NA	162	0.040	NA	1
C132	20	9.02×10^{-4}	398	403	528	38.0	NA	861	0.180	NA	0
C132	20	6.19×10^{-4}	432	491	NA	NA	NA	819	0.128	NA	0
C132	20	3.28×10^{-5}	396	461	549	33.0	65.0	850	0.152	NA	1
C132	20	1.02×10^{-3}	399	403	528	NA	NA	861	0.180	NA	0
C132	500	1.02×10^{-3}	297	341	349	16.0	NA	465	0.080	NA	1
C132	600	1.02×10^{-3}	192	215	212	NA	NA	257	0.047	NA	0
C132	650	1.02×10^{-3}	160	168	165	NA	NA	181	0.020	NA	1
C132	400	8.38×10^{-5}	339	423	483	32.0	NA	731	0.149	0.0079	1
C132	600	3.28×10^{-5}	138	190	190	NA	NA	223	0.085	0.0741	0
C132	600	8.40×10^{-5}	161	199	202	NA	NA	239	0.067	0.0634	1
C40	20	1.19×10^{-4}	436	495	572	33.0	63.0	831	0.128	0.0060	1
C40	20	1.19×10^{-4}	436	508	574	36.0	65.0	828	0.123	0.0080	0
C40	300	5.97×10^{-5}	379	481	544	27.0	53.0	847	0.149	0.0060	1
C40	431	5.97×10^{-5}	350	431	466	32.0	60.0	653	0.115	0.0070	1
C40	539	5.97×10^{-5}	230	296	305	28.0	41.0	392	0.095	0.0300	1
C40	647	5.97×10^{-5}	137	190	187	44.0	78.0	191	0.065	0.0830	1
C40	701	5.97×10^{-5}	89	127	125	55.0	86.0	123	0.054	0.0960	1
N8	20	1.19×10^{-4}	476	498	627	39.0	65.0	965	0.156	NA	0
N8	20	1.19×10^{-4}	471	486	634	39.0	68.0	987	0.159	NA	1
N8	300	5.97×10^{-5}	363	494	645	36.0	60.0	1096	0.204	0.0030	1
N8	431	5.97×10^{-5}	323	435	496	37.0	73.0	848	0.177	0.0050	1
N8	539	5.97×10^{-5}	233	321	340	45.0	80.0	465	0.121	0.0316	1
N8	647	5.97×10^{-5}	129	174	174	48.0	80.0	205	0.082	0.0590	1
N8	701	5.97×10^{-5}	86	122	123	54.0	84.0	150	0.101	0.0703	1
C10	20	1.19×10^{-4}	759	809	860	27.0	67.0	1091	0.074	0.0064	0
C10	20	1.19×10^{-4}	761	805	858	23.0	66.0	1095	0.076	0.0049	0

NA: not available; data not calculated

$S_y^{002} = S_y(0.002 \text{ offset})$: 0.2 % offset yield strength

$S_y^{02} = S_y(0.02 \text{ elong})$: stress measured at 2 % total strain

S_u = tensile strength; El_t = total elongation; RA = reduction of area

lm: used in calculation of stress-strain model. 1=used, 0=not used

continued on next page

Table 6: Data for WTC steels of this study.

Continued from the previous page

Specimen	T	$\dot{\epsilon}_0$	S_y^{002}	S_y^{02}	S_u	El_t	RA	K	n	m	lm
	°C	1/s	MPa	MPa	MPa	%	%	MPa			
C10	300	5.97×10^{-5}	589	751	801	24.0	60.0	1234	0.134	NA	0
C10	431	5.97×10^{-5}	613	708	712	23.0	64.0	946	0.080	0.0024	0
C10	539	5.97×10^{-5}	515	606	598	25.0	72.0	800	0.081	0.0203	0
C10	593	5.97×10^{-5}	343	506	498	30.0	79.0	1033	0.202	0.0423	0
C10	593	5.97×10^{-5}	378	508	498	29.0	79.0	966	0.185	0.0359	0
C10	593	5.97×10^{-5}	373	516	506	27.0	80.0	904	0.164	0.0385	0
C10	647	5.97×10^{-5}	181	353	356	37.0	76.0	662	0.227	0.1086	0
C10	701	5.97×10^{-5}	86	142	199	44.0	75.0	302	0.217	0.1574	0

NA: not available; data not calculated

$S_y^{002} = S_y(0.002 \text{ offset})$: 0.2 % offset yield strength

$S_y^{02} = S_y(0.02 \text{ elong})$: stress measured at 2 % total strain

S_u = tensile strength; El_t = total elongation; RA = reduction of area

lm: used in calculation of stress-strain model. 1=used, 0=not used

D.2 Stress-strain plots

Figure 15 plots the un-normalized true stress-strain behavior for the nine steels on identical axes. In each case, the curves are truncated at the tensile strength in the engineering stress-strain curve. Figures 16- 24 plot the measured true stress-strain data.

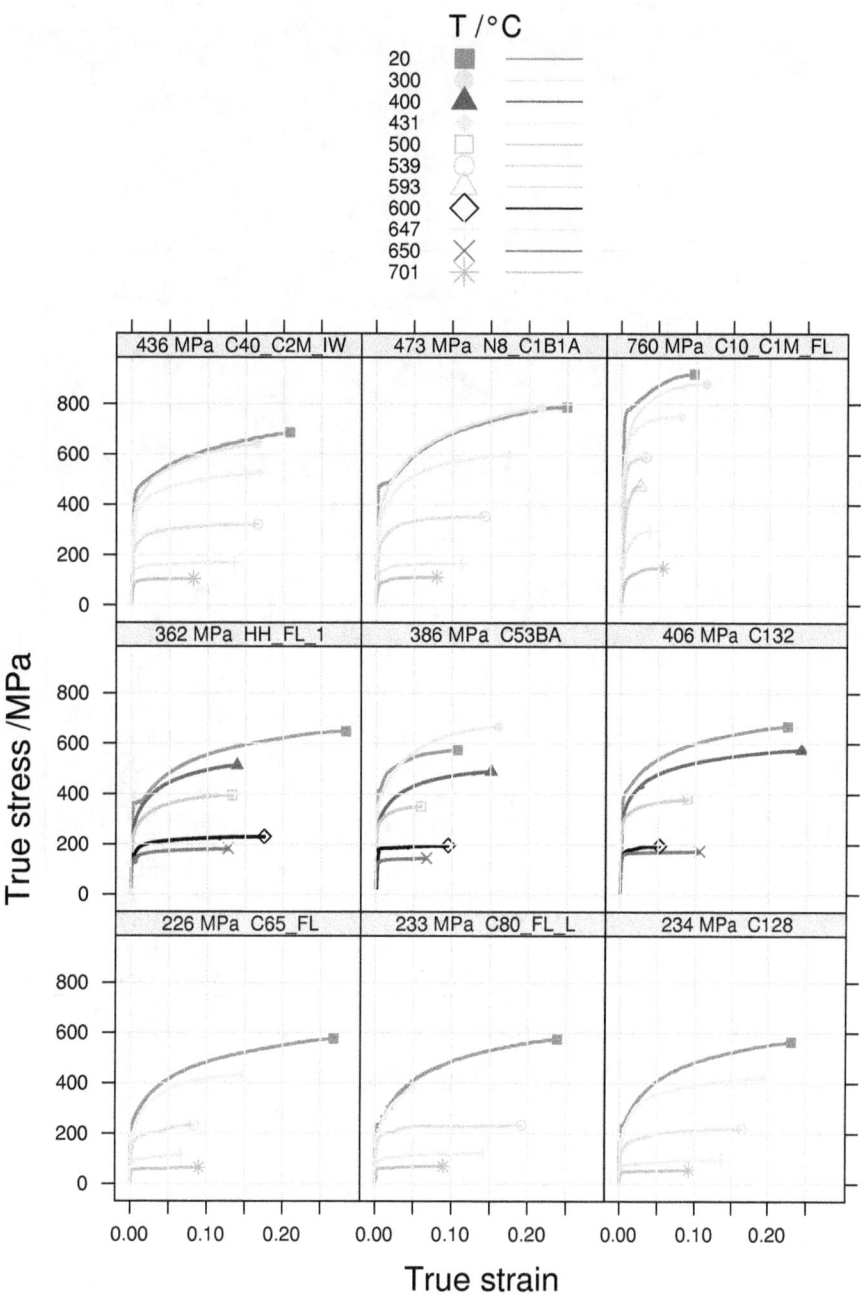

Figure 15: Stress-strain behavior of the nine steels.

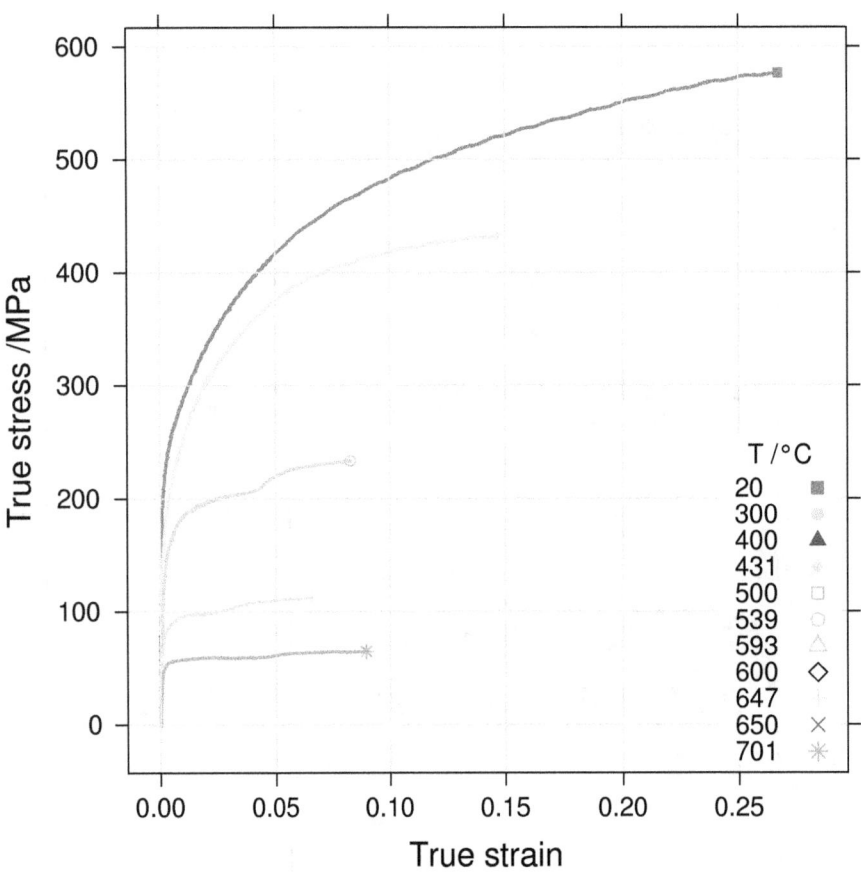

Figure 16: Stress-strain behavior of steel C65. $S_y(0.2\ \%\ \text{offset}) = 226$ MPa.

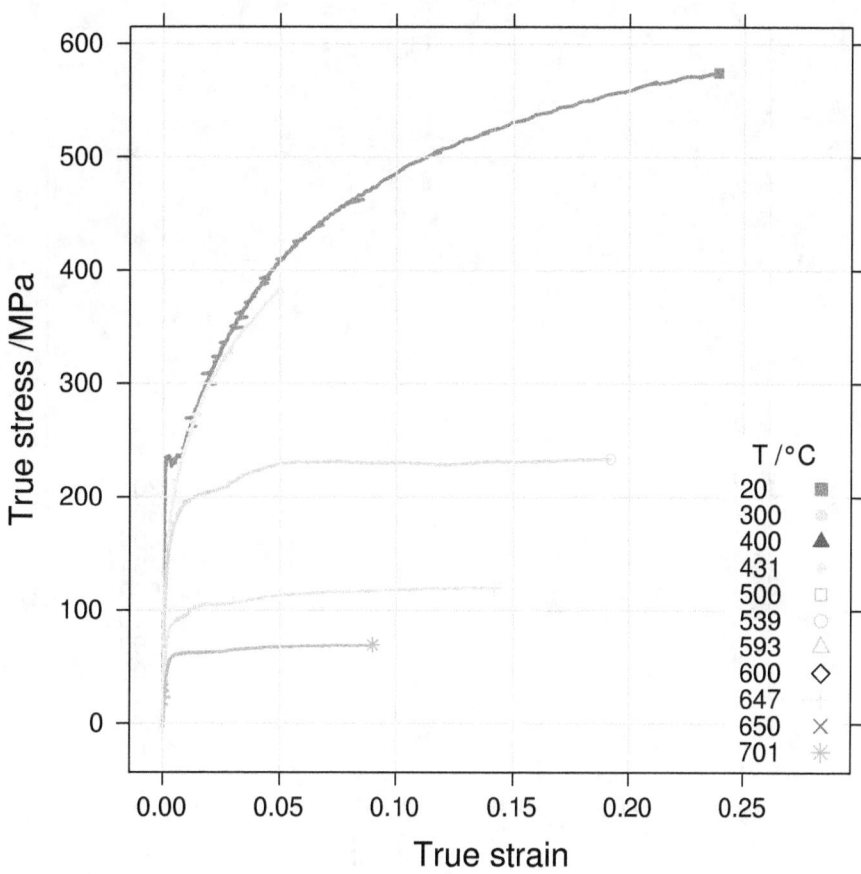

Figure 17: Stress-strain behavior of steel C80. $S_y(0.2\,\%\text{ offset}) = 233$ MPa.

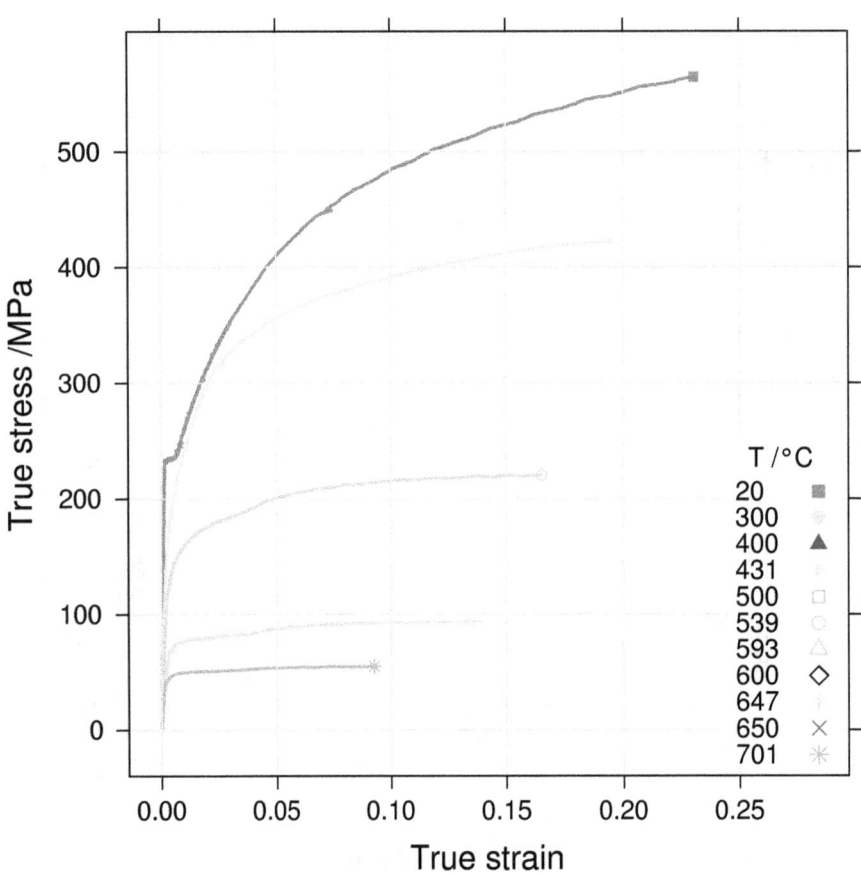

Figure 18: Stress-strain behavior of steel C128. $S_y(0.2 \% \text{ offset}) = 234$ MPa.

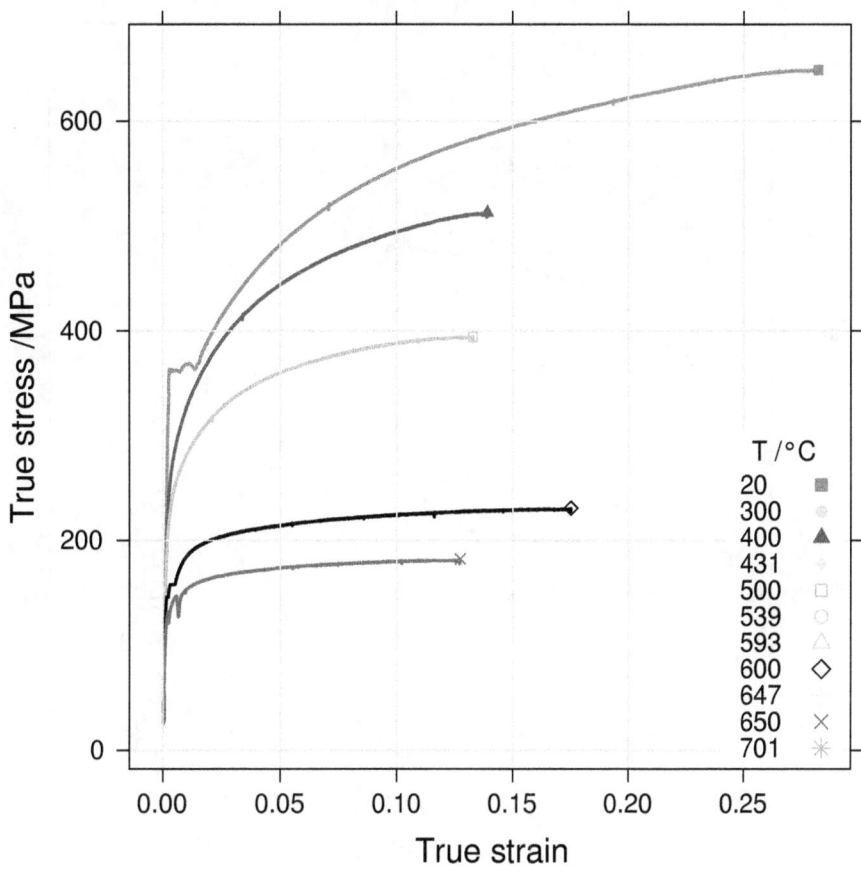

Figure 19: Stress-strain behavior of steel HH. $S_y(0.2 \% \text{ offset}) = 362.1$ MPa.

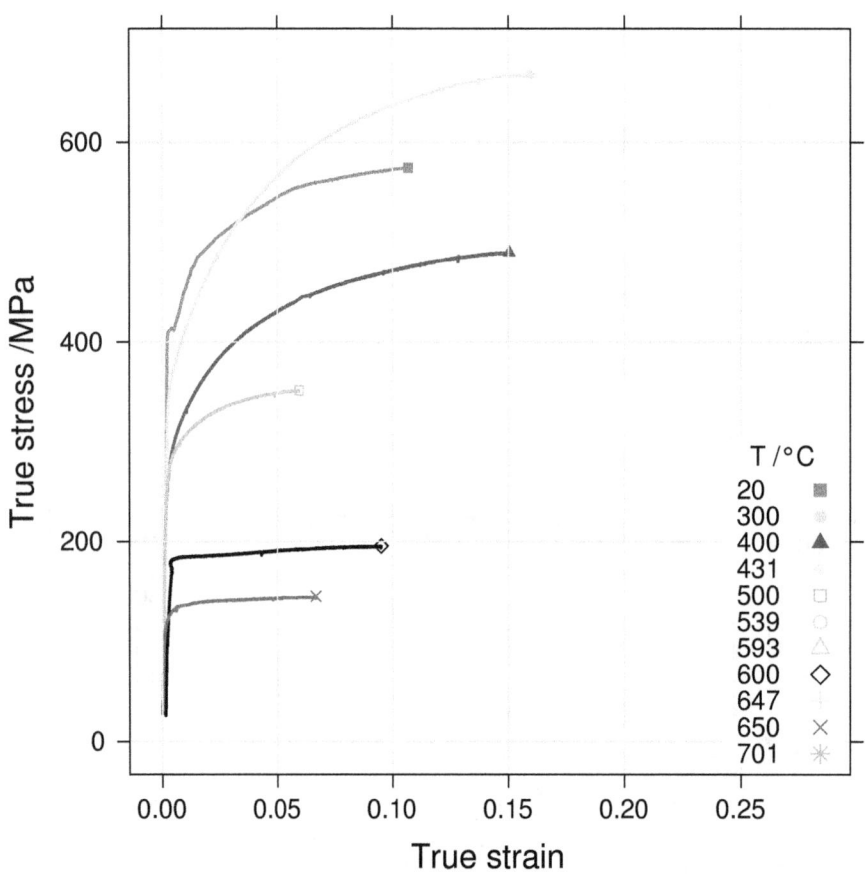

Figure 20: Stress-strain behavior of steel C53BA. $S_y(0.2\,\%\ \text{offset}) = 386$ MPa.

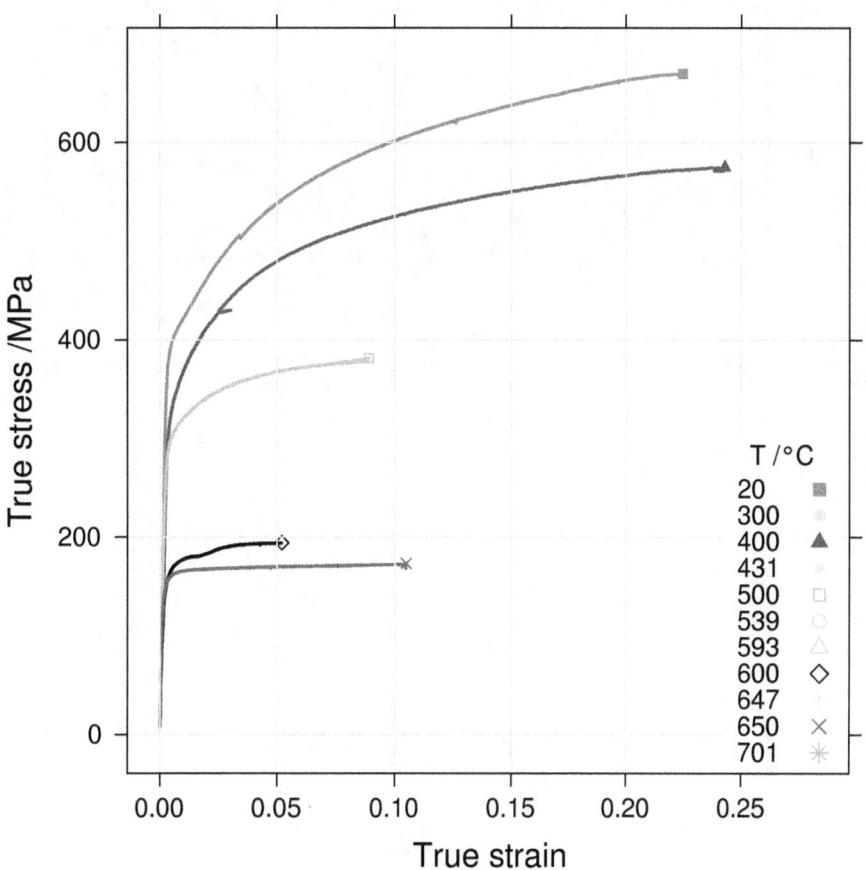

Figure 21: Stress-strain behavior of steel C132. $S_y(0.2 \% \text{ offset}) = 406$ MPa.

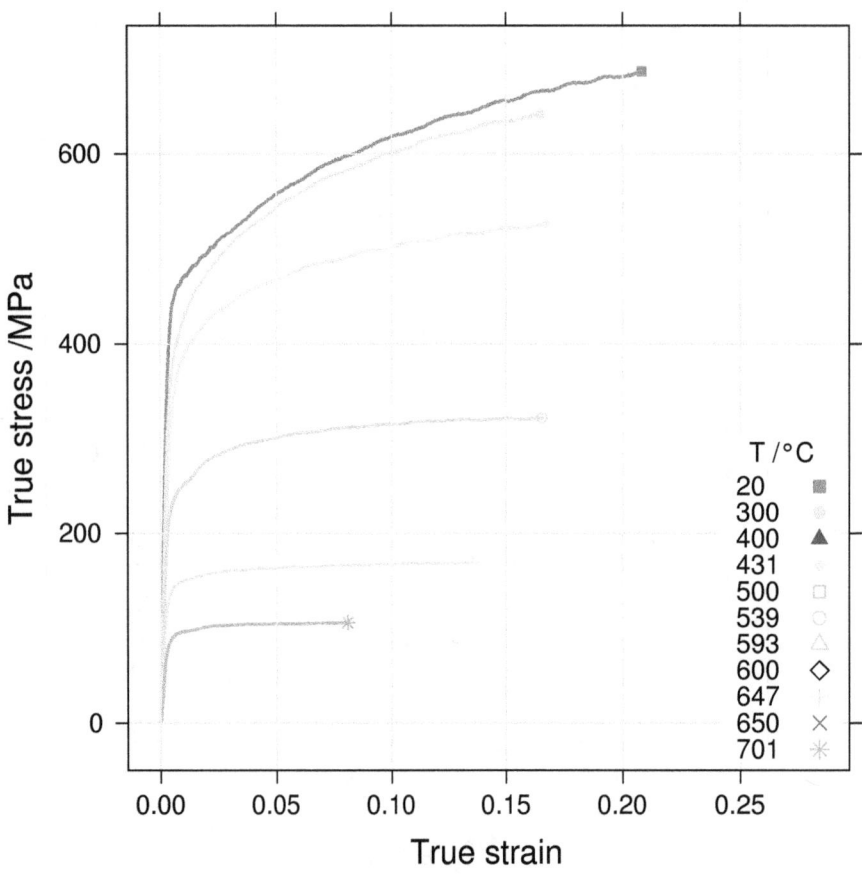

Figure 22: Stress-strain behavior of steel C40. $S_y(0.2\ \%\ \text{offset}) = 436$ MPa.

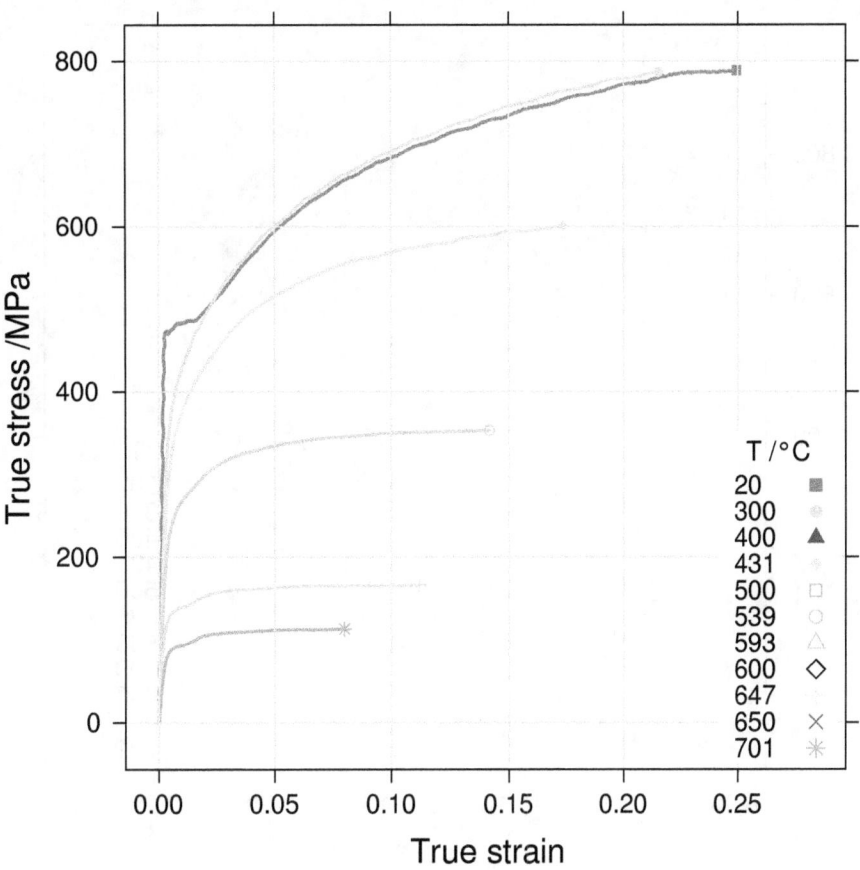

Figure 23: Stress-strain behavior of steel N8. $S_y(0.2\,\%\,\text{offset}) = 473.4\,\text{MPa}$.

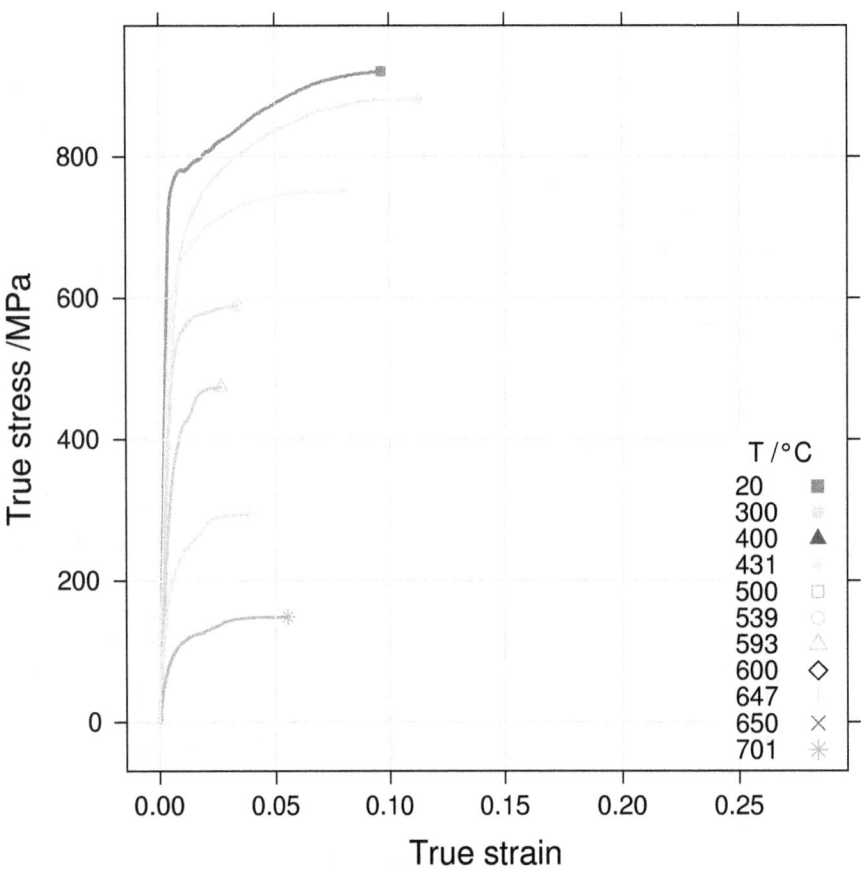

Figure 24: Stress-strain behavior of steel C10. $S_y(0.2 \% \text{ offset}) = 760$ MPa.

Table 7: Regression statistics for the fit of Eq. (1) to the retained strength data.

Parameter	Estimate	Standard Error	t
r_1	5.70839	1.29926	4.39357
r_2	1.00000	0.88480	1.13020
r_3	590.30863°C	42.56178°C	13.86945
r_4	918.72252°C	907.33326°C	1.01255

RSD: 0.08476 on 211 degrees of freedom

D.3 Retained Yield Strength Analysis

The parameters in Table 2 were calculated using the R statistical language [52], version 2.11, function `nls` with the "`port`" algorithm. Only data from Tables 6 and 10 with strain rates in the range were used. In addition the fit used only a subset of the data and was constrained:

Temperature range: $(300 \leq T \leq 800)$ °C,
Strain-rate range: $3.30 \times 10^{-5} < \dot{\epsilon} < 1.35 \times 10^{-4}$,
$1 \leq r_1 \leq 10$,
$1 \leq r_2 \leq 10$,
$200 \leq r_3 \leq 1500$,
$200 \leq r_4 \leq 1500$.

Table 7 summarizes the regression output. Constraining the values of $r_1 \geq 1$ and $r_2 \geq 1$ ensures that the shape of the retained strength does not become "S-shaped." Two of the parameters of the regression summarized in Table 7, r_2 and r_4 have large standard errors and correspondingly low values of t. In addition, the value of r_2 reached the lower limit of the constraint. Normally, this would be evidence that they should be omitted from the regression. However, using a two-parameter model produces very poor fidelity for temperatures, $T > 600$ °C, so the parameters have been retained.

The form of Figure 2 makes it difficult to assess the deviations of the data from the different fits. Figure 25 plots the difference between the prediction of the fit, $\hat{R}$ and the reported the retained strength, R, for each of the three fits shown in Figure 2. The solid lines in Figure 25 are a moving regression that represents the trend of the data. The fit of Eq. (1) is quite accurate up to 750 °C; it deviates by only about 1 % from the general trend of the data. Because the fit is required to asymptotically approach zero at high temperature, it systematically deviates from the data for the very highest temperatures. At these temperatures, the retained strength is extremely small, so the error is correspondingly small in absolute terms. Because

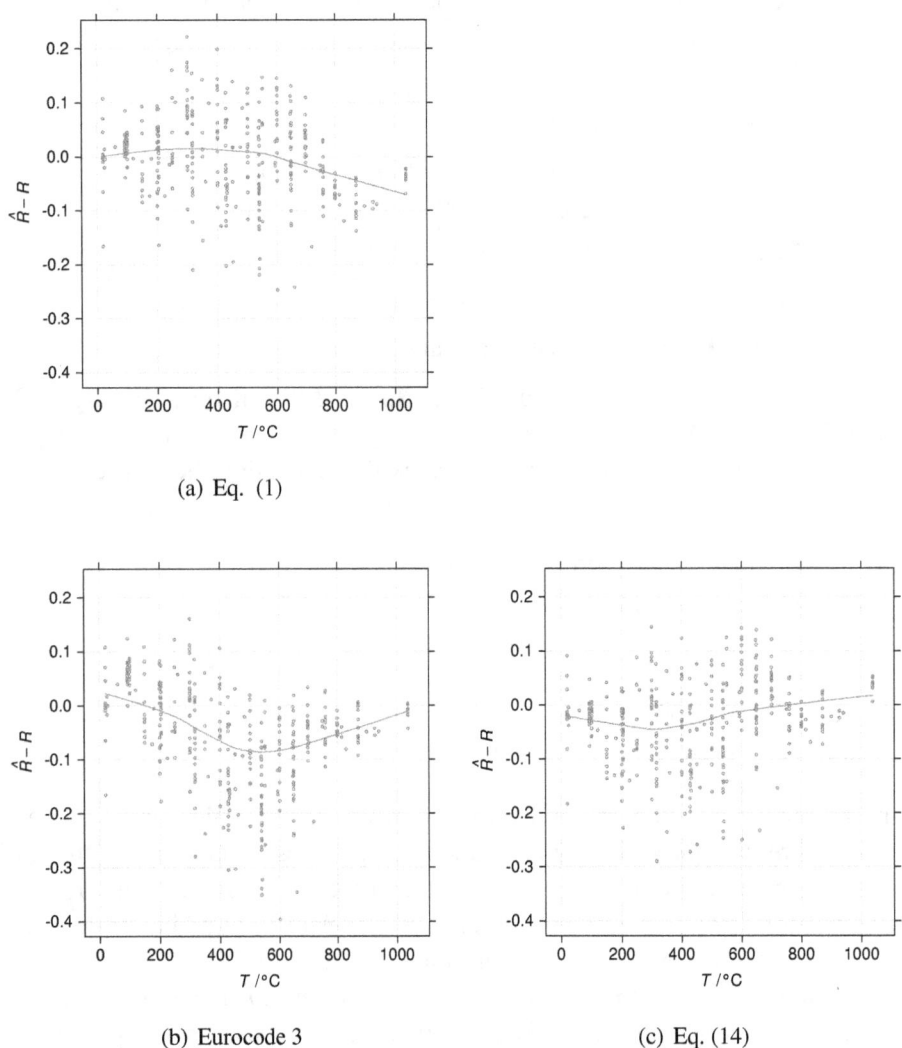

(a) Eq. (1)

(b) Eurocode 3

(c) Eq. (14)

Figure 25: Deviations from the fitted retained strength models. (a) This report, Equation (1) (b) Eurocode 3, see Appendix F.2 (c) Original WTC report, Eq. (14).

the value of R is used in calculating the stress-strain behavior, it is important to ensure fidelity across as much of the temperature range as possible. Figure 25b shows that the retained yield strength predicted by the Eurocode 3 formulation, see Appendix F.2, deviates significantly from the reported behavior in the mid-range of the temperatures. Figure 25c shows the retained yield strength behavior, R, predicted by the original World Trade Center report [12]

$$R = \frac{S_y}{S_y(T = 20°\text{C})} = A_2 + (1 - A_2) \exp\left(-\frac{1}{2}\left(\frac{T}{r_3}\right)^{r_1} - \frac{1}{2}\left(\frac{T}{r_4}\right)^{r_2}\right) \quad (14)$$

using the parameters of Table 2.

D.4   Calculation and analysis of strain rate sensitivity, m

D.4.1   Calculation

The high-temperature tensile tests to determine stress-strain behavior of this study were conducted according to ASTM E8 [10] and E21 [11]. These tensile test standards allow the user to increase the strain rate after yield. In strain-rate sensitive materials, increasing the strain rate increases the measured stress. The size of the stress jump can also be used to estimate the strain rate sensitivity, m from the change in initial and final strain rates:

$$m = \frac{\log_e \sigma_1 - \log_e \sigma_2}{\log_e \dot{\epsilon}_1 - \log_e \dot{\epsilon}_2} \quad (15)$$

where the subscripts "1" and "2" denote before and after the strain rate jump.

Figure 26 describes the procedure graphically. To estimate the the stresses, the center point of the stress jump in the stress-strain curve was identified graphically, and the stress-strain points in the transition region were deleted. The slopes of the stress strain curves were estimated by linear regression of a short section of stress-strain data before and after the jump. These linear fits were evaluated at the strain corresponding to the center of the transition region to calculate σ_1 and σ_2.

D.4.2   Analysis

To model the behavior of the strain-rate sensitivity of the individual steels as a function of temperature the parameters of Eq. (9) were fit using non-linear least squares method, subject to the constraints listed below.

m_0 not fit
$(1 \leq m_1 \leq 15)$,
$(100 \leq m_2 \leq 2000)$ °C,
$0.1 \leq m_3 \leq 0.2$
$T \geq 390$ °C

The value of m_0, which is effectively the strain rate sensitivity at room temperature was not fit. Instead, it was set to the mean value of the room-temperature measurements. In the case of specimen N8_C1B1A, where no room-temperature value existed, it was set to the average of all the room-temperature values. In addition, only data for $T \geq 390\ ^\circ\mathrm{C}$ was included in the fit to help ensure that the function evaluated to small values of m for low temperatures. Table 8 summarizes the results of the fits. The uncertainties of the parameter m_3 are large in several cases, even when the overall quality of the fit is good. They are high in these cases because the value of m_3 controls the limiting, high-temperature value of the strain-rate sensitivity. Steels whose data do not exhibit a limiting value will have large uncertainty associated with this parameter. Note that for the two steels, C128 and C65_FL, the value of $m_3 = 0.2$, which is the maximum allowed in the fit. Figure 27 shows the strain rate sensitivity for each steel, which is plotted in a single plot in the left-hand panel of Figure 5 as well. Insufficient data existed for several steels to compute a strain rate sensitivity, but Figure 27 contains entries for these steels to facilitate comparison with the other figures. The solid line in each panel is the fit of Eq. (9) with the parameters of Table 8.

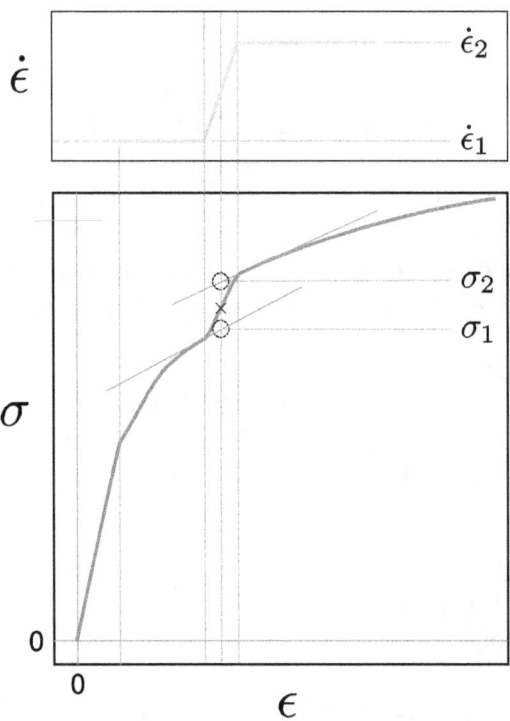

Figure 26: Illustration of the method for estimating the stress jump necessary to calculate the strain rate sensitivity.

Table 8: Parameters of fits plotted in Figure 27.

Specimen	m_0	m_1	m_2	m_3	RSD
			°C		
C65_FL	0.007583	8.1±1.3	645.1±25.5	0.142±0.022	0.004
C80_FL_L	0.004700	6.4±2.9	632.0±95.8	0.158±0.073	0.013
C128	0.008800	3.2±5.4	699.9±1392.3	0.200±0.750	0.027
HH_FL_1	NA	$NA \pm NA$	$NA \pm NA$	$NA \pm NA$	NA
C53BA	NA	$NA \pm NA$	$NA \pm NA$	$NA \pm NA$	NA
C132	NA	$NA \pm NA$	$NA \pm NA$	$NA \pm NA$	NA
C40_C2M_IW	0.007000	10.4±1.0	608.6±7.8	0.090±0.004	0.003
N8_C1B1A	0.006646	8.2±3.4	599.6±40.5	0.064±0.012	0.007
C10_C1M_FL	0.005650	15.0±1.6	648.9±7.4	0.159±0.010	0.005

NA: not available, data not calculated

D.5 Calculation of the strain-hardening parameters

The five parameters for the strain-hardening model, k_1, k_2, k_3, k_4, and n, Eq. (10), were calculated by a non-linear least-squares regression of the entire set of forty-three stress-strain curves comprising about 32 000 individual points, using the R statistical language [52], version 2.11, function `nls` with the "`port`" algorithm. The parameters of the fit were constained:

Temperature range: $(300 \leq T \leq 800)$ °C,
Strain-rate range: $3.30 \times 10^{-5} < \dot{\epsilon} < 1.35 \times 10^{-4}$,
$1 \leq k_1 \leq 10$,
$(100 \leq k_2 \leq 1000)$ °C
$(100 \leq k_3 \leq 2000)$ MPa,
$200 \leq k_4 \leq 1500$.
$200 \leq n \leq 1500$.

The fit did not approach the constraints. The individual curves used are identified in Table 6 by the entry "lm." The data set excluded the $F_y = 100$ ksi steel (C10) because this steel is not common in building construction. Although Table 6 reports all the tests, some of which replicated conditions, only one temperature condition was used for each specimen, to avoid biasing the results toward one specific condition. The individual stress-strain curves did not, however, contain equal numbers of stress-strain pairs, so a small bias may exist toward the stress-strain curves with the most individual points. In fitting the parameters, the strain data were limited

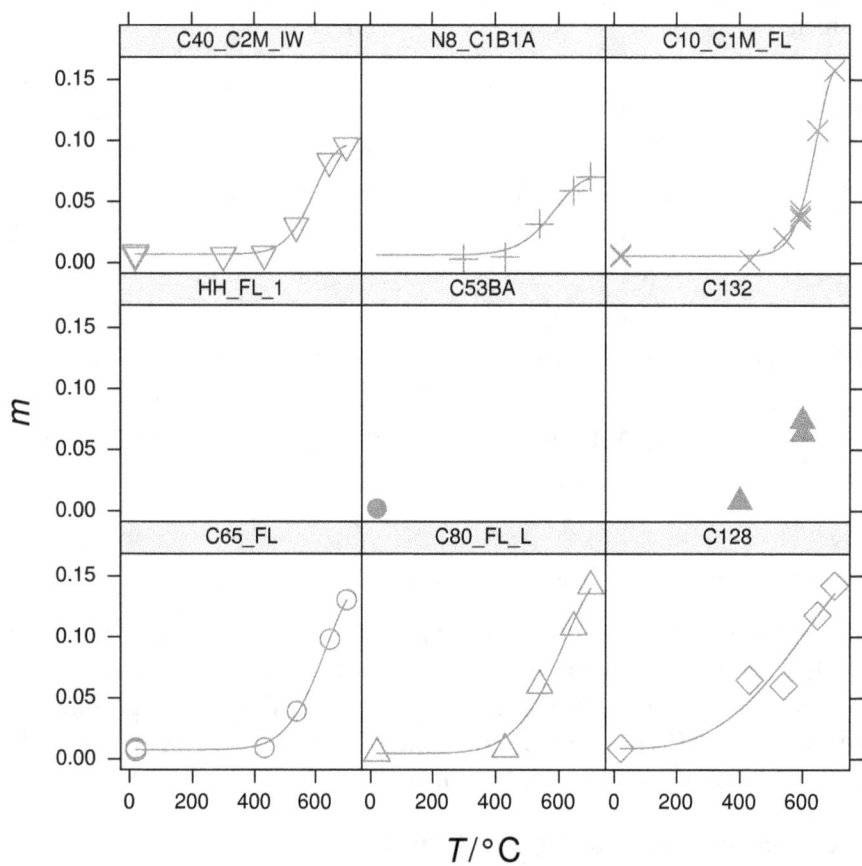

Figure 27: Dependence of the strain-rate sensitivity, m, on temperature. Solid lines are the fits of Eq. (9).

to the range ($0.01 < \epsilon < 0.15$. When the maximum strain was less than the upper limit, the maximum strain was used. Choosing this upper limit constrained the fit to regions where the stress-strain data was appropriate to fire modeling. The retained strength, R, Eq. (1) was computed for each temperature using the values in Table 3. The measured room-temperature yield strength, S_y^0, was used for each the steel. The effect of strain rate was included via the expression for the strain rate sensitivity, Eq. (9), with the parameters for the m_i summarized in Table 3. Table 9 summarizes the output of that regression. Note that the elastic modulus is not relevant in the regression, because the strain range only includes plastic strain.

Parameter	Estimate	Standard Error	t
k_1	8.29390	0.07215	114.95378
k_2	537.64846	0.42718	1258.59509
k_3	959.43380	7.74224	123.92201
k_4	0.76632	0.01457	52.59578
n	0.48337	0.00212	227.71218

RSD: 30.09051 on 32252 degrees of freedom

Table 9: Output from the regression to determine the strain-hardening parameters for Eq (7).

Figure 28 corresponds to Figure 6, except that it plots the absolute deviation, D_a

$$D_a = \hat{\sigma} - \sigma \tag{16}$$

of the global fit to the steel data, Eq. (7). The model is only evaluated for the plastic strain: $\epsilon > S_y/E$. The largest deviation is about 100 MPa, and it is typically less than 50 MPa.

The large absolute deviations at small strain are caused by the difference between the elastic behavior as modeled, and the low-strain behavior in the individual tests. The poorest overall agreement of the model with the data is around $T = 300\,^{\circ}\text{C}$. This region is also where the spread in the retained yield strength, R, is the largest, Figure 2. It is also the region in which dynamic strain aging (DSA) effects are most prevalent in steels [53, 54].

E   Literature Data

E.1   Descriptions of steels from the literature used in this report

This report uses data assembled from many literature sources. In many cases the data were taken from tables in the individual references, but in others they were digitized directly from plots. This section summarizes the data sources. Table 10 compiles all the literature data used in creating the plots and conclusions of this report. This section also summarizes *some* of the other notable studies [5, 13, 21, 55–57] on high-temperature stress-strain behavior of structural steel that do not appear in plots or Table 10, typically because they were obtained at unconventional strain rates.

E.1.1   Brockenbrough

Year: 1968
Reference: [58]

Steel description Data for ASTM A36, and four grades of United States Steel structural steels: T1, COR-TEN, TRI-TEN, and MAN-TEN (ASTM A440). T1 was a quenched-and-tempered construction steel with $F_y = 100$ ksi, supplied to ASTM A514, "High-Yield-Strength, Quenched and Tempered Alloy Steel Plate, Suitable for Welding." COR-TEN steels were intended as steels with improved atmospheric corrosion resistance, and could be supplied to ASTM A242 " High-Strength Low-Alloy Structural Steel" and ASTM A588 " High-Strength Low-Alloy Structural Steel, up to 50 ksi [345 MPa] Minimum Yield Point, with Atmospheric Corrosion Resistance." TRI-TEN was a high-strength low-alloy steel with

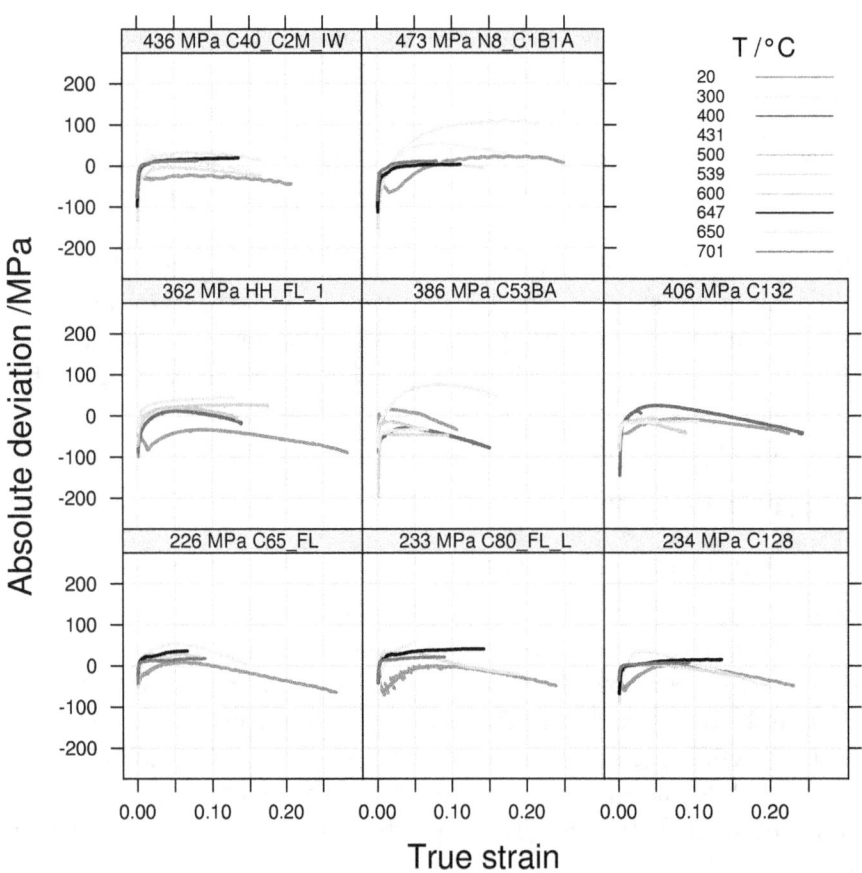

Figure 28: Absolute deviation, D_a between true stress predicted by Eq. (7), $\hat{\sigma}$, and the measured stress, σ.

F_y ==50 ksi, that was produced to ASTM A441 "High-strength low alloy Structural Manganese Vanadium Steel." That standard was withdrawn and replaced by ASTM A572 in 1988. ASTM A440, "High Strength Structural Steel" was another F_y = 50 ksi steel intended to "provide high strength at a price slightly less than that of TRI-TEN steel" [58] was withdrawn in 1979 without replacement.

Notes Data for the USS grades appear in Holt [26, 27], Sections E.1.8 and E.1.9.

E.1.2 Borvik

Year: 2005
Reference: [55]
T **range:** $(20 \leq T \leq 500)$ °C
Chemistry: C 0.08 Si 0.26 Mn 1.40 S 0.002 P 0.008 Nb 0.028 V 0.04 Ti 0.01 Cr 0.02 Ni 0.04 Mo 0.007 N 0.006 Al 0.032

Steel description Weldox 460 E is a thermomechanically rolled, weldable, fine-grained $t = 12$ mm plate with V, Nb, Ti. The minimum F_y = 460 MPa.

Notes The manuscript describes tests to establish impact properties, but contains low strain-rate data. High-rate tests were conducted in a split-Hopkinson bar test machine. The strain-rate sensitivity, m, is based on the measured stress σ at $\epsilon = 0.02$. Figure 8 in the manuscript [55] shows stress-strain behavior at $\dot{\epsilon} = 5 \times 10^{-4}$ 1/s. The report only presents data for the effect of strain rate on stress measured at $e = 0.02$, and not for the effect of strain rate on the yield strength, S_y. It shows no stress-strain curves for low-rate tests.

E.1.3 Chen

Year: 2006
Reference: [23]
T **range:** $(22 \leq T \leq 940)$ °C
Chemistry: C 0.22 Si 0.55 Mn 1.7 P 0.04 S 0.03 Cr 0.30 Ni 0.5 Cu 0.4

Steel description Chen et al. describe the steel as "XLERPLATE Grade 350" which is delivered to Australian AS3678 ("Structural Steel–Hot-rolled plates, floor plates, and slabs"). Specimens were fabricated from $t = 5$ mm plate.

Notes The manuscript contains test results, but not stress-strain curves, for high-temperature tensile tests conducted at $\dot{\epsilon} = 1 \times 10^{-4}$ 1/s. It also contains results temperature-ramp tests, in which the load is held constant while the temperature is increased. In addition to the data for AS3678, the manuscript also reports data for a high-strength steel, with S_y(0.002 offset)=780 MPa. The manuscript reports strength reductions for other yield strengths, up to $e = 0.02$.

E.1.4   Chijiiwa

Year: 1993
Reference: [24]
T **range:** $(20 \leq T \leq 700)\,°C$
Chemistry: NA

Steel description   The conventional steel was a Japanese SM490A steel plate. SM490A is delivered to JIS G 3106 "Rolled steel for welded structures." The "490" in SM490A refers to the minimum tensile strength in MPa. The minimum yield strength is $F_y = 325$ MPa. JIS G 3106 specifies composition limits for C, Si, Mn, P, and S, and "an alloying element other than the above-mentioned can be added as needed."

Notes   The manuscript compares conventional SM490A steel to fire-resistive steel designed to meet the SM490A standard. In addition to high-temperature tensile test results, the manuscript also contains creep data and microstructural analysis. The fire-resistive steel tensile tests were conducted to the Japanese high-temperature tensile test standard [37] JIS G 0567, which prescribes a strain rate, $\dot{\epsilon} = 0.00005$ 1/s. Presumably the conventional steel tests were also conducted to that standard. The manuscript contains only plots of retained strength, and does not show stress-strain curves.

E.1.5   Fujimoto

Year: 1980
Reference: [56]
T **range:** $(20 \leq T \leq 600)\,°C$
Chemistry: C 0.18 Mn 0.73 Si 0.24 P 0.016 S 0.017

Steel description   The steel is a Japanese SS41, $t = 25$ mm plate. SS41 is structural steel plate delivered to Standard JIS 3101, "Rolled Steels for general structure." The "41" refers to the minimum tensile strength measured in kg/mm^2. The minimum yield point is $F_y = 285$ MPa. A recent version of JIS G 3101 set composition limits for P and S only.

Notes   Figure 5 in the manuscript shows tensile stress-strain curves conducted to $\epsilon = 0.03$ in the range, but lacks evidence for any specific strain rate.

E.1.6   Gowda

Year: 1978
Reference: [25]
T **range:** $(25 \leq T \leq 927)\,°C$
Chemistry: C 0.13 Mn 0.82 Si 0.21 P 0.014 S 0.006

Steel description The steel was ASME SA 516 pressure vessel plate for moderate-temperature service from a $t = 9.5$ mm plate. This steel is specified to have a minimum yield strength $F_y = 207$ MPa. ASME SA 516 is equivalent to ASTM A516, "Standard specification for Pressure Vessel Plates, Carbon Steel, for Moderate- and Lower-Temperature Service."

Notes The strain rate sensitivity calculated was based on the tensile strength, TS. The manuscript contains a long data table of all the tensile test results as well as stress-strain curves for low-strain and higher strain as Figures 2 and 3.

E.1.7 Harmathy

Year: 1970
Reference: [13]
T **range:** $(20 \leq T \leq 650)\,°\mathrm{C}$
Chemistry: A36: C 0.19 Mn 0.71 Si 0.09 P 0.007 S 0.03
Chemistry: CSA G40.12: C 0.195 Mn 1.4 Si 0.022 Cu 0.08 Ni 0.08 Cr 0.01

Steel description The ASTM A36 is described as Si semi-killed, open-hearth, hot-rolled $t = 12.7$ mm. The Canadian CSA G 40.12 is described as Si semi-killed, basic oxygen process, hot-rolled $t = 12.7$ mm plate. Canadian standard CSA G 40.12, "General Purpose Structural Steel," has a specified yield strength $F_y = 44$ ksi. [59].

Notes The manuscript contains graphs of full stress-strain curves (Figure 3) as well as creep curves. Tensile tests were conducted at a single strain rate, which is higher than typical high-temperature tensile test rates. In addition, the manuscript contains data for ASTM A 421 pre-stressing steel, at three different strain rates.

E.1.8 Holt

Year: Unknown
Reference: [27]
T **range:** $(27 \leq T \leq 1038)\,°\mathrm{C}$
Chemistry: C 0.17 Mn 0.86 Cu 0.03 Ni 0.03 Cr 0.46 Mo 0.19 V 0.040

Steel description USS T-1A, ASTM A514 quenched and tempered. See further description under Brockenbrough, Section E.1.1.

Notes The manuscript is an undated technical report from United States Steel, Corp. Tests were conducted according to ASTM E21-58T, so $\dot{\epsilon} = 8.33 \times 10^{-5}$ 1/s. The manuscript also contains data for USS T1, which appear in Brockenbrough [58], Figure 1.5.

E.1.9 Holt

Year: 1964
Reference: [26]
T **range:** $(27 \leq T \leq 1038)\,°C$
Chemistry: A36: C 0.24 Mn 0.96 P 0.011 S 0.027 Si 0.043 Cu 0.058
Chemistry: Cor-ten $t = 70$ mm: C 0.17 Mn 1.00 P 0.018 S 0.030 Si 0.22 Ni 0.022
Cr 0.48 V 0.042 Cu 0.3
Chemistry: Tri-ten $t = 12.7$ mm: C 0.20 Mn 1.09 P 0.054 S 0.025 Si 0.025 V 0.038
Cu 0.26 N 0.005
Chemistry: Tri-ten $t = 25$ mm: C 0.20 Mn 1.16 P 0.016 S 0.035 Si 0.04 Ni 0.015
Cr 0.035 N 0.006 V 0.056 Cu 0.22
Chemistry: Tri-Ten $t = 50$ mm: C 0.21 Mn 1.25 P 0.017 S 0.028 Si 0.09 Ni 0.028
Cr 0.064 N 0.006 V 0.056 Cu 0.25
Chemistry: A440 $t = 12.7$ mm: C 0.29 Mn 1.54 P 0.026 S 0.038 Si 0.05 Cu 0.31
Chemistry: A440 $t = 50$ mm: C 0.20 Mn 1.60 P 0.017 S 0.018 Si 0.20 N 0.006
Cu 0.23
Chemistry: A440 $t = 25$ mm: C 0.26 Mn 1.48 P 0.017 S 0.036 Si 0.035 N 0.007
Cu 0.24

Steel description The steels were plates of different thicknesses, t, of USS Cor-Ten, Tri-Ten and ASTM A440 and ASTM A36 steels. Cor-ten is a weathering steel. See other description under Brockenbrough, Section E.1.1.

Notes This manuscript is an unpublished US Steel technical report. Much of the data in this USS Technical Report appear in Spaeder [34], Sec. E.1.22, and were identified by comparing the chemistry and plate thicknesses. In addition, Brockenbrough [58], Sec. E.1.1, Fig.1.5 shows the average values for the steels. The report shows only retained strengths, and no stress-strain curves.

E.1.10 Hu

Year: 2009
Reference: [40]
T **range:** $(20 \leq T \leq 800)\,°C$
Chemistry: NA

Steel description The steel was taken from the web of a W30x99 shape supplied to ASTM A992 ('Standard Specification for Structural Steel Shapes") structural steel.

Notes The manuscript contains the stress-strain behavior out to failure as well as expanded views up to $e = 0.1$. The tests employed two rates, both of which are greater than the standard ASTM E21 test rate. Tests also included determination of the static yield strength via a load-relaxation method.

E.1.11 Jerath

Year: 1980
Reference: [5]
T **range:** $(20 \leq T \leq 800)\,°\mathrm{C}$
Chemistry: see manuscript

Steel description The manuscript describes fifteen different structural steels produced by the British Steel and supplied to BS4360 ("Weldable Structural Steels.") Grades 43A, 43E, 50B, and 50D, as well as several other steels.

Notes This report does not appear in any library. Tests were conducted at a strain rate $\dot{\epsilon} = 2.78 \times 10^{-4}$ 1/s, which is higher than the usual testing rate. For each steel, a page describes the chemistry, and tabulates the retained yield and tensile strength, the reduction of area and elongation for temperatures up to 800 °C and sometimes 1000 °C. No stress-strain curves are shown.

E.1.12 Kirby

Year: 1988
Reference: [18]
T **range:** $(20 \leq T \leq 700)\,°\mathrm{C}$
Chemistry: NA

Steel description British BS4360 Grade 50B from flanges of notch-tough structural sections. Specification BS4360, "Weldable Structural Steels" was the common structural steel specification in Britain at the time.

Notes The manuscript contains a figure that shows retained strength determined from tensile tests, denoted as "steady-state conditions." The manuscript also contains data for constant-load, ramping-temperature tests, denoted as "transient state conditions." The tensile tests were conducted according to the British standard BS3688, which was superseded by BS EN 10002-5:1992 "Tensile testing of metallic materials. Method of test at elevated temperatures." That standard re-quires $\dot{\epsilon} = 5 \times 10^{-5}$ 1/s.

E.1.13 Kirby

Year: 1993
Reference: [60]
T **range:** $(20 \leq T \leq 800)\,°\mathrm{C}$
Chemistry: BS15 9J/428/H $t = 14$ mm: C 0.14 Si 0.04 Mn 0.70 P 0.017 S 0.060 Cr 0.03 Ni 0.09 Cu 0.19 N 0.0046 Sn 0.087

Chemistry: BS15 OJ055/JFP $t = 15.9$ mm: C 0.21 Si 0.04 Mn 0.53 P 0.021 S 0.031 Cr 0.02 Ni 0.04 Cu 0.03 N 0.0028

Chemistry: BS15 OJ/072/J $t = 25.4$ mm: C 0.22 Si 0.04 Mn 0.53 P 0.055 S 0.037 Cr 0.03 Ni 0.04 Cu 0.02 N 0.0025

Chemistry: BS4360Gr43A NA

Steel description The manuscript reports data from three historical steels taken from a structure built in 1939, and presumably supplied to BS15:1936 "Standard Specification for Structural Steel for Bridges and General Building Construction," which only specified the tensile strength. These three steels were complemented by three more modern steels supplied to BS4360:Grade 43A "Weldable Structural Steels." The three BS4360 steels were taken from 73 kg/m, 240 kg/m, and 54 kg/m rolled sections.

Notes The data are also published in Ref. 28. Tests were conducted according to the British standard BS3688. which was superseded by BS EN 10002-5:1992 "Tensile testing of metallic materials. Method of test at elevated temperatures." That test requires $\dot{\epsilon} = 5 \times 10^{-5}$ 1/s. The manuscript contains plots of various strength measures as a function of temperature for the six steels. However,it is not explicit on which steel is assigned which symbol in the figures. I have assigned them based on their order in the data tables. No stress-strain curves are shown.

E.1.14 Li

Year: 2003
Reference: [21]
T **range:** $(20 \leq T \leq 700)\,°C$
Chemistry: NA

Steel description The manuscript describes the steel as "constructional steel widely used in China... used in steel frameworks."

Notes Tests were conducted at $\dot{\epsilon} = 6.7 \times 10^{-4}$ 1/s, which is about ten times higher than rates specified in the usual high-temperature tensile standards. Even at this high rate, the retained strength is somewhat low. The authors report an expression for the retained yield and tensile strength that is only valid in the temperature range of the tests, because it decreased to below zero at high temperature. The manuscript also contains information for a low-carbon steel used for bolts.

E.1.15 Lou

Year: 1995
Reference: [29,61]
T **range:** $(20 \leq T \leq 300)\,°C$

Chemistry: CSA G40.21 350 WT plate $t = 12.7$ mm C 0.19 Si 0.17 Mn 1.07 P 0.006 S 0.010 Cr 0.02 Ni 0.02 Cu 0.06 N 0.0042 V<0.01

Chemistry: CSA G40.21 350 AT plate $t = 17.8$ mm C 0.12 Si 0.22 Mn 1.12 P 0.007 S 0.009 Cr 0.42 Ni 0.37 Cu 0.27 N 0.0059 V 0.046

Steel description The authors described the steels as produced in fine-grain condition using fully killed steelmaking practice. The plate was supplied to Canadian standard G40.21, "General Requirements for Rolled or Welded Structural Quality Steel/ Structural Quality Steel" The "W" denotes weldable. The "A" denotes a high-strength-low-alloy steel with improved corrosion resistance. The "T" indicates that Charpy tests for toughness are required. The "350" denotes the yield strength in units of MPa.

Notes Tensile tests were conducted over a two-decade strain rate range.

E.1.16 Manjoine

Year: 1944
Reference: [62]
T **range:** $(20 \leq T \leq 600)\,°C$

Steel description The manuscript describes the steel as "mild steel," a commercial low-carbon open hearth steel annealed 1 h at 920 °C in dissociated ammonia.

Notes The manuscript contains stress-strain curves for the entire temperature range at different strain rates, as well as graphical representation of the decrease in strength with temperature. The data in included here primarily for historic value, since this was one of the first determinations of the rate dependence of steel strength.

E.1.17 Outinen

Year: 2001
Reference: [30]
T **range:** $(20 \leq T \leq 950)\,°C$
Chemistry: S355 NA

Steel description The steel is described as S355, which is a European EN SFS-En 10 025(1993) hotrolled $t = 4$ mm sheet. The report contains no microstructure or chemistry information for the steel.

Notes Most of the yield and tensile strength data presented in the report, Tables 6 and 8, are derived from the so-called transient state tests, in which the specimen is heated at constant temperature ramp rate under fixed load. The tensile tests reported in Figure 12 are also summarized in Appendix 2 as distinct stress-strain points. All the true tensile tests were conducted to EN 10 002-5, which prescribes a strain rate, $\dot{\epsilon} = 5 \times 10^{-5}$ 1/s, though the report does not explicitly state this rate. The report also contains data for steel from structural steel tubes and cold-rolled steel sheet. The data calculated from temperature-ramp test, called "transient state tests," also were published in 1997 [19], and are not included here.

E.1.18   Poh

Year: 1998
Reference: [31]
T **range:** $(20 \leq T \leq 800)$ °C
Chemistry: C 0.16 P 0.018 Mn 1.46 Si 0.15 S 0.006 Ni 0.003 Cr 0.009 Cu 0.013
              Sn 0.002 Mo 0.003 V 0.004 Ti 0.001 Nb 0.001 Al 0.024

Steel description   The steel, from a heavy, hot-rolled shape, was supplied to Australian standard AS3679.1 Grade 300, "Hot-rolled structural steel bars and sections." Kotwal [63] describes Australian steel specifications in detail. Grade 300 designates the nominal specified minimum yield strength measured in MPa.

Notes   The report is a very detailed Ph. D. thesis. It contains stress-strain curves for the steels. Some of the material was later published in 2001 [45]. The 2001 paper contains a very detailed, 42-parameter model for representing the stress-strain curve as a function of temperature.

E.1.19   Sakumoto

Year: 1999
Reference: [32]
T **range:** $(20 \leq T \leq 800)$ °C
Chemistry: NA

Steel description   The steel was supplied to Japan JIS 3106 SM400A, hot-rolled steel. JIS Standard 3106 is titled "Rolled steel for welded structures." Grade SM 400A refers to the minimum tensile strength measured in N/mm^2. The specified minimum yield strength is $F_y = 245$ N/mm^2. Takanashi [64] describes the Japanese structural steel specifications in detail.

Notes   The manuscript contains stress-strain curves up to $e = 0.04$ as a function of temperature for steels SM400A, fire-resistive steel NSFR400A, and SUS304, an austenitic stainless steel. The testing rate $\dot{\epsilon} = 5 \times 10^{-5}$ 1/s is inferred from the citation of Japanese tensile test standard, JIS G 0567 [37].

E.1.20   Skinner

Year: 1973

66

Reference: [33]

T **range:** $(20 \leq T \leq 650)\,°\mathrm{C}$

Chemistry: NA

Steel description The steel was supplied to Australian AS A186:250 steel "Structural Steels (Ordinary Weldable Grades." Kotwal [63] describes the Australian steel grades in greater detail, and indicates that A186:250 was intended to be similar to ASTM A36. The "250" refers to the nominal specified yield strength, F_y, measured in MPa.

Notes The manuscript contains much of the information that appears in a BHP report [6]. Skinner includes includes data for for five different structural steels, but only stress-strain curves for the A186 steel. The manuscript also contains graphic representation of the decrease in strength with temperature, the effect of strain rate on strength and temperature, creep curves, and constant-load temperature-ramp test data.

The actual values for stress and strain in Figure 11 of this report were taken from the tables in the BHP report [6]

E.1.21   Smith, ASTM DS11

Year: 1970

Reference: [2]

T **range:** $(21 \leq T \leq 1038)\,°\mathrm{C}$

Chemistry: see manuscript

Steel description This ASTM report contains data for 37 different plate materials for pressure vessels, supplied to ASTM Standards A201, A212, A285, A299, A442, A515, and A516.

Notes The report is a supplement to ASTM STP 180 [65], often called DS 11. Relatively little data are reported for $T > 538\,°\mathrm{C}$. It also contains extensive data tables that summarize the chemistry and processing of the steels, but does not show any stress-strain curves. Because the report is essentially a table of data, it is unethical to simply reprint the table here. Figure 29 replots the data from the report on the same scale with the same overlays as Figure 2. Although the fit to Eq. (1) represents the data reasonably well, unlike the structural steel data, some of the data drop below the $R = 0.5$ line at 538 °C. The report also contains a similar quantity of data obtained on the same grades made into tubes, as well as some data from wrought bar. None of the data from the report are used in the analysis of this report, however.

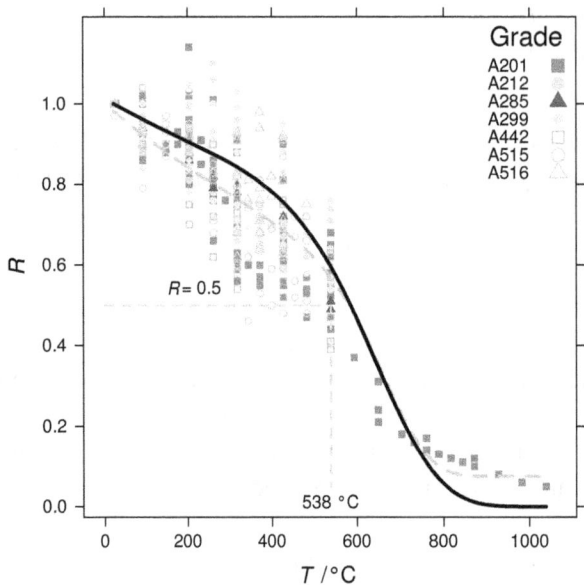

Figure 29: Normalized, retained yield strength, R, as a function of temperature for different ASTM grades of pressure vessel steels reported in Ref. [2] Table III-P. The strength is the 0.2 % offset yield strength, $S_y(0.002$ offset) normalized to its room-temperature value. Solid lines have the same meaning as in Figure 2.

E.1.22 Spaeder

Year: 1977
Reference: [34]
T **range:** $(27 \leq T \leq 649)\,°C$
Chemistry: see manuscript

Steel description   The manuscript contains data for eight different USS Cor-ten A and Cor-ten B weathering steels. Both production and experimental heats are included.

Notes   Some data from this paper also appear also appear in Holt [26]. The manuscript contains excellent data tables for test results and chemistry, but no stress-strain curves. The identification of the plate thicknesses allowed correlation with the data in Holt [26]. Strain rates were assigned based on reference to ASTM E21 [11], $\dot{\epsilon} = 8.33 \times 10^{-5}$ 1/s. .

E.1.23 Stevens

Year: 1971
Reference: [6]
T **range:** $(27 \leq T \leq 650)$ °C
Chemistry: NA

Steel description The report contains extensive data for four Australian structural steels: AS A186:250 L0, AS A186:250, AS187:WR350/1, and AS A186:400 L15. Kotwal [63] describes Standard AS A186 "Structural Steel-Ordinary Weldable Grades" as an omnibus specification similar to ASTM A36. The numerical designations 250 and 400 refer to the minimum yield stress measured in MPa, rounded to the nearest 50 MPa. The designations "L0" and "L15" referred to additional Charpy test requirements for $T = 0$ °C and $T = -15$ °C. AS A187, "Structural Steels-Weather-resistant Weldable Grades," was a specification for weathering steels supplied by BHP as their "AUS-TEN" series. Again, the designation 350 referred to the yield strength in MPa, rounded to the nearest 50 MPa. The "/1" identified the lower carbon variant: maximum C=0.12 % mass fraction.

Notes Information about the steels was correlated by comparing the steel descriptions to those in Skinner [33]. This report contains data for AS A186:250 steel that Skinner [33] published later. Tables of stress and fixed strain points are included for all four steels.

E.1.24 United States Steel 1972 elevated temperature handbook

Year: 1968
Reference: [35]
T **range:** A36: $(27 \leq T \leq 704)$ °C
Chemistry: A36: NA
T **range:** A514: $(27 \leq T \leq 1038)$ °C
Chemistry: A514: C 0.14 Mn 0.76 Cu 0.3 Ni 0.86 Cr 0.55 Mo 0.45 V 0.03

Steel description ASTM A36 Steel, from United States Steel. It is not the same A36 steel that appears in Holt [26] and Brockenbrough [58], see Sections E.1.9 and E.1.1, based on examination of retained strengths and temperature increments. The A514 steel is USS T1 Steel, quenched-and-tempered steel plate. Its chemistry was not specified in the report, but was assigned based on comparison with data from Holt [27], which contains graphical and tabular data, but not stress-strain curves.

Notes The introduction to the report states that tests were conducted to ASTM E21, so $\dot\epsilon = 8.33 \times 10^{-5}$ 1/s. The report contains retained strengths, and not stress-strain curves.

E.2 Tabular data for the literature structural steels

Table 10 summarizes the mechanical property data taken from the literature sources, that appear in Figures 2, 7, 5, and 36.

Table 10: Data for structural steels from literature sources.

Reference	Section	Steel	T	$\dot{\epsilon}$	S_y^{002}	S_y^{02}
			°C	1/s	MPa	MPa
[23]	E.1.3	AS3678 XLERPlate	22	1.00×10^{-4}	401	465
[23]	E.1.3	AS3678 XLERPlate	60	1.00×10^{-4}	385	446
[23]	E.1.3	AS3678 XLERPlate	120	1.00×10^{-4}	381	446
[23]	E.1.3	AS3678 XLERPlate	150	1.00×10^{-4}	377	446
[23]	E.1.3	AS3678 XLERPlate	180	1.00×10^{-4}	369	446
[23]	E.1.3	AS3678 XLERPlate	240	1.00×10^{-4}	361	488
[23]	E.1.3	AS3678 XLERPlate	300	1.00×10^{-4}	381	516
[23]	E.1.3	AS3678 XLERPlate	360	1.00×10^{-4}	345	484
[23]	E.1.3	AS3678 XLERPlate	410	1.00×10^{-4}	361	479
[23]	E.1.3	AS3678 XLERPlate	460	1.00×10^{-4}	325	432
[23]	E.1.3	AS3678 XLERPlate	540	1.00×10^{-4}	313	400
[23]	E.1.3	AS3678 XLERPlate	600	1.00×10^{-4}	285	344
[23]	E.1.3	AS3678 XLERPlate	660	1.00×10^{-4}	225	256
[23]	E.1.3	AS3678 XLERPlate	720	1.00×10^{-4}	140	144
[23]	E.1.3	AS3678 XLERPlate	830	1.00×10^{-4}	60	60
[23]	E.1.3	AS3678 XLERPlate	940	1.00×10^{-4}	36	40
[24]	E.1.4	SMA490	20	5.00×10^{-5}	334	NA
[24]	E.1.4	SMA490	100	5.00×10^{-5}	321	NA
[24]	E.1.4	SMA490	200	5.00×10^{-5}	271	NA
[24]	E.1.4	SMA490	300	5.00×10^{-5}	229	NA
[24]	E.1.4	SMA490	400	5.00×10^{-5}	217	NA
[24]	E.1.4	SMA490	500	5.00×10^{-5}	188	NA
[24]	E.1.4	SMA490	600	5.00×10^{-5}	117	NA
[24]	E.1.4	SMA490	700	5.00×10^{-5}	56	NA
[25]	E.1.6	ASME-SA516	25	1.00×10^{-4}	307	319
[25]	E.1.6	ASME-SA516	25	1.00×10^{-4}	297	NA

NA: not available; data not calculated

$S_y^{002} = S_y(0.002 \text{ offset})$: 0.2 % offset yield strength

$S_y^{02} = S_y(0.02 \text{ elong})$: stress measured at 2 % total strain

continued on next page

Table 10: Data for structural steels from literature sources.

Continued from the previous page

Reference	Section	Steel	T	$\dot{\epsilon}$	S_y^{002}	S_y^{02}
			°C	1/s	MPa	MPa
[25]	E.1.6	ASME-SA516	315	1.00×10^{-4}	208	328
[25]	E.1.6	ASME-SA516	315	1.00×10^{-4}	208	NA
[25]	E.1.6	ASME-SA516	427	1.00×10^{-4}	192	289
[25]	E.1.6	ASME-SA516	427	1.00×10^{-4}	192	NA
[25]	E.1.6	ASME-SA516	537	1.00×10^{-4}	164	215
[25]	E.1.6	ASME-SA516	537	1.00×10^{-4}	169	NA
[25]	E.1.6	ASME-SA516	649	1.00×10^{-4}	107	121
[25]	E.1.6	ASME-SA516	649	1.00×10^{-4}	108	NA
[25]	E.1.6	ASME-SA516	760	1.00×10^{-4}	38	41
[25]	E.1.6	ASME-SA516	760	1.00×10^{-4}	40	NA
[25]	E.1.6	ASME-SA516	815	1.00×10^{-4}	34	NA
[25]	E.1.6	ASME-SA516	815	1.00×10^{-4}	40	NA
[25]	E.1.6	ASME-SA516	871	1.00×10^{-4}	35	NA
[25]	E.1.6	ASME-SA516	871	1.00×10^{-4}	38	NA
[25]	E.1.6	ASME-SA516	927	1.00×10^{-4}	26	NA
[25]	E.1.6	ASME-SA516	927	1.00×10^{-4}	30	NA
[25]	E.1.6	ASME-SA516	25	5.00×10^{-3}	308	NA
[25]	E.1.6	ASME-SA516	315	5.00×10^{-3}	221	NA
[25]	E.1.6	ASME-SA516	427	5.00×10^{-3}	205	NA
[25]	E.1.6	ASME-SA516	537	5.00×10^{-3}	179	NA
[25]	E.1.6	ASME-SA516	537	5.00×10^{-3}	187	NA
[25]	E.1.6	ASME-SA516	649	5.00×10^{-3}	138	NA
[25]	E.1.6	ASME-SA516	649	5.00×10^{-3}	141	NA
[25]	E.1.6	ASME-SA516	760	5.00×10^{-3}	62	NA
[25]	E.1.6	ASME-SA516	760	5.00×10^{-3}	66	NA
[25]	E.1.6	ASME-SA516	871	5.00×10^{-3}	49	NA
[25]	E.1.6	ASME-SA516	871	5.00×10^{-3}	52	NA
[25]	E.1.6	ASME-SA516	25	1.00×10^{-1}	339	NA
[25]	E.1.6	ASME-SA516	315	1.00×10^{-1}	244	NA
[25]	E.1.6	ASME-SA516	427	1.00×10^{-1}	204	NA

NA: not available; data not calculated

$S_y^{002} = S_y(0.002 \text{ offset})$: 0.2 % offset yield strength

$S_y^{02} = S_y(0.02 \text{ elong})$: stress measured at 2 % total strain

continued on next page

Table 10: Data for structural steels from literature sources.

Continued from the previous page

Reference	Section	Steel	T	$\dot{\epsilon}$	S_y^{002}	S_y^{02}
			°C	1/s	MPa	MPa
[25]	E.1.6	ASME-SA516	537	1.00×10^{-1}	198	NA
[25]	E.1.6	ASME-SA516	537	1.00×10^{-1}	200	NA
[25]	E.1.6	ASME-SA516	649	1.00×10^{-1}	166	NA
[25]	E.1.6	ASME-SA516	649	1.00×10^{-1}	169	NA
[25]	E.1.6	ASME-SA516	760	1.00×10^{-1}	82	NA
[25]	E.1.6	ASME-SA516	760	1.00×10^{-1}	83	NA
[25]	E.1.6	ASME-SA516	871	1.00×10^{-1}	67	NA
[25]	E.1.6	ASME-SA516	871	1.00×10^{-1}	67	NA
[26]	E.1.9	A36	27	8.33×10^{-5}	248	NA
[26]	E.1.9	A36	93	8.33×10^{-5}	228	NA
[26]	E.1.9	A36	204	8.33×10^{-5}	239	NA
[26]	E.1.9	A36	316	8.33×10^{-5}	216	NA
[26]	E.1.9	A36	427	8.33×10^{-5}	219	NA
[26]	E.1.9	A36	538	8.33×10^{-5}	181	NA
[26]	E.1.9	A36	649	8.33×10^{-5}	90	NA
[26]	E.1.9	A36	760	8.33×10^{-5}	50	NA
[26]	E.1.9	A36	871	8.33×10^{-5}	37	NA
[26]	E.1.9	A36	1038	8.33×10^{-5}	11	NA
[26]	E.1.9	A440	27	8.33×10^{-5}	414	NA
[26]	E.1.9	A440	93	8.33×10^{-5}	414	NA
[26]	E.1.9	A440	204	8.33×10^{-5}	442	NA
[26]	E.1.9	A440	316	8.33×10^{-5}	436	NA
[26]	E.1.9	A440	427	8.33×10^{-5}	396	NA
[26]	E.1.9	A440	538	8.33×10^{-5}	332	NA
[26]	E.1.9	A440	649	8.33×10^{-5}	185	NA
[26]	E.1.9	A440	760	8.33×10^{-5}	61	NA
[26]	E.1.9	A440	871	8.33×10^{-5}	27	NA
[26]	E.1.9	A440	1038	8.33×10^{-5}	9	NA
[26]	E.1.9	A440	27	8.33×10^{-5}	341	NA
[26]	E.1.9	A440	93	8.33×10^{-5}	318	NA

NA: not available; data not calculated

$S_y^{002} = S_y(0.002 \text{ offset})$: 0.2 % offset yield strength

$S_y^{02} = S_y(0.02 \text{ elong})$: stress measured at 2 % total strain

continued on next page

72

Table 10: Data for structural steels from literature sources.

Continued from the previous page

Reference	Section	Steel	T	$\dot{\epsilon}$	S_y^{002}	S_y^{02}
			°C	1/s	MPa	MPa
[26]	E.1.9	A440	204	8.33×10^{-5}	309	NA
[26]	E.1.9	A440	316	8.33×10^{-5}	310	NA
[26]	E.1.9	A440	427	8.33×10^{-5}	276	NA
[26]	E.1.9	A440	538	8.33×10^{-5}	247	NA
[26]	E.1.9	A440	649	8.33×10^{-5}	112	NA
[26]	E.1.9	A440	760	8.33×10^{-5}	37	NA
[26]	E.1.9	A440	871	8.33×10^{-5}	17	NA
[26]	E.1.9	A440	1038	8.33×10^{-5}	11	NA
[26]	E.1.9	A440	27	8.33×10^{-5}	334	NA
[26]	E.1.9	A440	93	8.33×10^{-5}	313	NA
[26]	E.1.9	A440	204	8.33×10^{-5}	314	NA
[26]	E.1.9	A440	316	8.33×10^{-5}	322	NA
[26]	E.1.9	A440	427	8.33×10^{-5}	292	NA
[26]	E.1.9	A440	538	8.33×10^{-5}	272	NA
[26]	E.1.9	A440	649	8.33×10^{-5}	157	NA
[26]	E.1.9	A440	760	8.33×10^{-5}	73	NA
[26]	E.1.9	A440	871	8.33×10^{-5}	42	NA
[26]	E.1.9	A440	1038	8.33×10^{-5}	14	NA
[26]	E.1.9	Cor-Ten	27	8.33×10^{-5}	354	NA
[26]	E.1.9	Cor-Ten	760	8.33×10^{-5}	62	NA
[26]	E.1.9	Cor-Ten	871	8.33×10^{-5}	43	NA
[26]	E.1.9	Cor-Ten	1038	8.33×10^{-5}	13	NA
[26]	E.1.9	Cor-Ten	27	8.33×10^{-5}	318	NA
[26]	E.1.9	Cor-Ten	93	8.33×10^{-5}	302	NA
[26]	E.1.9	Cor-Ten	204	8.33×10^{-5}	300	NA
[26]	E.1.9	Cor-Ten	316	8.33×10^{-5}	291	NA
[26]	E.1.9	Cor-Ten	427	8.33×10^{-5}	273	NA
[26]	E.1.9	Cor-Ten	538	8.33×10^{-5}	221	NA
[26]	E.1.9	Cor-Ten	649	8.33×10^{-5}	130	NA
[26]	E.1.9	Cor-Ten	760	8.33×10^{-5}	54	NA

NA: not available; data not calculated

$S_y^{002} = S_y(0.002 \text{ offset})$: 0.2 % offset yield strength

$S_y^{02} = S_y(0.02 \text{ elong})$: stress measured at 2 % total strain

continued on next page

Table 10: Data for structural steels from literature sources.

Continued from the previous page

Reference	Section	Steel	T	$\dot{\epsilon}$	S_y^{002}	S_y^{02}
			°C	1/s	MPa	MPa
[26]	E.1.9	Cor-Ten	871	8.33×10^{-5}	36	NA
[26]	E.1.9	Cor-Ten	1038	8.33×10^{-5}	22	NA
[26]	E.1.9	Tri-Ten	27	8.33×10^{-5}	364	NA
[26]	E.1.9	Tri-Ten	93	8.33×10^{-5}	319	NA
[26]	E.1.9	Tri-Ten	204	8.33×10^{-5}	297	NA
[26]	E.1.9	Tri-Ten	316	8.33×10^{-5}	267	NA
[26]	E.1.9	Tri-Ten	427	8.33×10^{-5}	252	NA
[26]	E.1.9	Tri-Ten	538	8.33×10^{-5}	205	NA
[26]	E.1.9	Tri-Ten	649	8.33×10^{-5}	121	NA
[26]	E.1.9	Tri-Ten	760	8.33×10^{-5}	50	NA
[26]	E.1.9	Tri-Ten	871	8.33×10^{-5}	30	NA
[26]	E.1.9	Tri-Ten	1038	8.33×10^{-5}	12	NA
[26]	E.1.9	Tri-Ten	27	8.33×10^{-5}	325	NA
[26]	E.1.9	Tri-Ten	93	8.33×10^{-5}	312	NA
[26]	E.1.9	Tri-Ten	204	8.33×10^{-5}	281	NA
[26]	E.1.9	Tri-Ten	316	8.33×10^{-5}	278	NA
[26]	E.1.9	Tri-Ten	427	8.33×10^{-5}	254	NA
[26]	E.1.9	Tri-Ten	538	8.33×10^{-5}	192	NA
[26]	E.1.9	Tri-Ten	649	8.33×10^{-5}	126	NA
[26]	E.1.9	Tri-Ten	760	8.33×10^{-5}	58	NA
[26]	E.1.9	Tri-Ten	871	8.33×10^{-5}	32	NA
[26]	E.1.9	Tri-Ten	1038	8.33×10^{-5}	11	NA
[26]	E.1.9	Tri-Ten	27	8.33×10^{-5}	321	NA
[26]	E.1.9	Tri-Ten	93	8.33×10^{-5}	306	NA
[26]	E.1.9	Tri-Ten	204	8.33×10^{-5}	288	NA
[26]	E.1.9	Tri-Ten	316	8.33×10^{-5}	283	NA
[26]	E.1.9	Tri-Ten	427	8.33×10^{-5}	268	NA
[26]	E.1.9	Tri-Ten	538	8.33×10^{-5}	212	NA
[26]	E.1.9	Tri-Ten	649	8.33×10^{-5}	128	NA
[26]	E.1.9	Tri-Ten	760	8.33×10^{-5}	55	NA

NA: not available; data not calculated

$S_y^{002} = S_y(0.002 \text{ offset})$: 0.2 % offset yield strength

$S_y^{02} = S_y(0.02 \text{ elong})$: stress measured at 2 % total strain

continued on next page

74

Table 10: Data for structural steels from literature sources.

Continued from the previous page

Reference	Section	Steel	T	$\dot{\epsilon}$	S_y^{002}	S_y^{02}
			°C	1/s	MPa	MPa
[26]	E.1.9	Tri-Ten	871	8.33×10^{-5}	31	NA
[26]	E.1.9	Tri-Ten	1038	8.33×10^{-5}	12	NA
[27]	E.1.8	USS T1a	27	8.30×10^{-5}	816	NA
[27]	E.1.8	USS T1a	93	8.30×10^{-5}	779	NA
[27]	E.1.8	USS T1a	204	8.30×10^{-5}	723	NA
[27]	E.1.8	USS T1a	316	8.30×10^{-5}	671	NA
[27]	E.1.8	USS T1a	427	8.30×10^{-5}	634	NA
[27]	E.1.8	USS T1a	538	8.30×10^{-5}	514	NA
[27]	E.1.8	USS T1a	649	8.30×10^{-5}	198	NA
[27]	E.1.8	USS T1a	760	8.30×10^{-5}	66	NA
[27]	E.1.8	USS T1a	871	8.30×10^{-5}	43	NA
[27]	E.1.8	USS T1a	1038	8.30×10^{-5}	18	NA
[18]	E.1.12	BS4360 Gr 50B	20	5.00×10^{-5}	395	422
[18]	E.1.12	BS4360 Gr 50B	100	5.00×10^{-5}	367	401
[18]	E.1.12	BS4360 Gr 50B	200	5.00×10^{-5}	336	422
[18]	E.1.12	BS4360 Gr 50B	300	5.00×10^{-5}	300	414
[18]	E.1.12	BS4360 Gr 50B	400	5.00×10^{-5}	284	389
[18]	E.1.12	BS4360 Gr 50B	500	5.00×10^{-5}	239	300
[18]	E.1.12	BS4360 Gr 50B	600	5.00×10^{-5}	160	186
[18]	E.1.12	BS4360 Gr 50B	700	5.00×10^{-5}	83	97
[28]	E.1.13	BS15 9J/428/J (dt)	20	5.00×10^{-5}	258	259
[28]	E.1.13	BS15 9J/428/J (dt)	100	5.00×10^{-5}	235	281
[28]	E.1.13	BS15 9J/428/J (dt)	150	5.00×10^{-5}	250	316
[28]	E.1.13	BS15 9J/428/J (dt)	200	5.00×10^{-5}	257	323
[28]	E.1.13	BS15 9J/428/J (dt)	250	5.00×10^{-5}	198	NA
[28]	E.1.13	BS15 9J/428/J (dt)	300	5.00×10^{-5}	192	288
[28]	E.1.13	BS15 9J/428/J (dt)	400	5.00×10^{-5}	189	NA
[28]	E.1.13	BS15 9J/428/J (dt)	500	5.00×10^{-5}	167	215
[28]	E.1.13	BS15 9J/428/J (dt)	600	5.00×10^{-5}	94	109
[28]	E.1.13	BS15 9J/428/J (dt)	700	5.00×10^{-5}	48	47

NA: not available; data not calculated

$S_y^{002} = S_y(0.002 \text{ offset})$: 0.2 % offset yield strength

$S_y^{02} = S_y(0.02 \text{ elong})$: stress measured at 2 % total strain

continued on next page

Table 10: Data for structural steels from literature sources.

Continued from the previous page

Reference	Section	Steel	T	$\dot{\epsilon}$	S_y^{002}	S_y^{02}
			°C	1/s	MPa	MPa
[28]	E.1.13	BS15 9J/428/J (dt)	800	5.00×10^{-5}	29	32
[28]	E.1.13	BS15- OJ/055/JFP (ob)	20	5.00×10^{-5}	230	274
[28]	E.1.13	BS15- OJ/055/JFP (ob)	100	5.00×10^{-5}	218	287
[28]	E.1.13	BS15- OJ/055/JFP (ob)	150	5.00×10^{-5}	221	324
[28]	E.1.13	BS15- OJ/055/JFP (ob)	200	5.00×10^{-5}	219	335
[28]	E.1.13	BS15- OJ/055/JFP (ob)	250	5.00×10^{-5}	166	320
[28]	E.1.13	BS15- OJ/055/JFP (ob)	300	5.00×10^{-5}	159	301
[28]	E.1.13	BS15- OJ/055/JFP (ob)	400	5.00×10^{-5}	151	270
[28]	E.1.13	BS15- OJ/055/JFP (ob)	500	5.00×10^{-5}	136	186
[28]	E.1.13	BS15- OJ/055/JFP (ob)	600	5.00×10^{-5}	78	92
[28]	E.1.13	BS15- OJ/055/JFP (ob)	700	5.00×10^{-5}	40	43
[28]	E.1.13	BS15- OJ/055/JFP (ob)	800	5.00×10^{-5}	29	32
[28]	E.1.13	BS15 OJ/072/J (ut)	20	5.00×10^{-5}	217	314
[28]	E.1.13	BS15 OJ/072/J (ut)	100	5.00×10^{-5}	201	319
[28]	E.1.13	BS15 OJ/072/J (ut)	150	5.00×10^{-5}	218	379
[28]	E.1.13	BS15 OJ/072/J (ut)	175	5.00×10^{-5}	215	377
[28]	E.1.13	BS15 OJ/072/J (ut)	200	5.00×10^{-5}	211	368
[28]	E.1.13	BS15 OJ/072/J (ut)	225	5.00×10^{-5}	210	372
[28]	E.1.13	BS15 OJ/072/J (ut)	250	5.00×10^{-5}	191	354
[28]	E.1.13	BS15 OJ/072/J (ut)	300	5.00×10^{-5}	175	342
[28]	E.1.13	BS15 OJ/072/J (ut)	400	5.00×10^{-5}	167	317
[28]	E.1.13	BS15 OJ/072/J (ut)	500	5.00×10^{-5}	152	230
[28]	E.1.13	BS15 OJ/072/J (ut)	600	5.00×10^{-5}	94	110
[28]	E.1.13	BS15 OJ/072/J (ut)	700	5.00×10^{-5}	49	48
[28]	E.1.13	BS15 OJ/072/J (ut)	800	5.00×10^{-5}	29	32
[28]	E.1.13	BS436043A-1	20	5.00×10^{-5}	326	326
[28]	E.1.13	BS436043A-1	100	5.00×10^{-5}	306	309
[28]	E.1.13	BS436043A-1	150	5.00×10^{-5}	331	351
[28]	E.1.13	BS436043A-1	200	5.00×10^{-5}	333	416
[28]	E.1.13	BS436043A-1	250	5.00×10^{-5}	290	348

NA: not available; data not calculated

$S_y^{002} = S_y(0.002 \text{ offset})$: 0.2 % offset yield strength

$S_y^{02} = S_y(0.02 \text{ elong})$: stress measured at 2 % total strain

continued on next page

Table 10: Data for structural steels from literature sources.

Continued from the previous page

Reference	Section	Steel	T	$\dot{\epsilon}$	S_y^{002}	S_y^{02}
			°C	1/s	MPa	MPa
[28]	E.1.13	BS436043A-1	300	5.00×10^{-5}	247	339
[28]	E.1.13	BS436043A-1	400	5.00×10^{-5}	223	341
[28]	E.1.13	BS436043A-1	500	5.00×10^{-5}	209	260
[28]	E.1.13	BS436043A-1	600	5.00×10^{-5}	110	125
[28]	E.1.13	BS436043A-1	700	5.00×10^{-5}	48	48
[28]	E.1.13	BS436043A-1	800	5.00×10^{-5}	29	32
[28]	E.1.13	BS436043A-2	20	5.00×10^{-5}	284	331
[28]	E.1.13	BS436043A-2	100	5.00×10^{-5}	278	359
[28]	E.1.13	BS436043A-2	150	5.00×10^{-5}	283	396
[28]	E.1.13	BS436043A-2	200	5.00×10^{-5}	284	402
[28]	E.1.13	BS436043A-2	250	5.00×10^{-5}	267	398
[28]	E.1.13	BS436043A-2	300	5.00×10^{-5}	220	362
[28]	E.1.13	BS436043A-2	400	5.00×10^{-5}	213	333
[28]	E.1.13	BS436043A-2	500	5.00×10^{-5}	198	275
[28]	E.1.13	BS436043A-2	600	5.00×10^{-5}	140	145
[28]	E.1.13	BS436043A-2	700	5.00×10^{-5}	53	52
[28]	E.1.13	BS436043A-2	800	5.00×10^{-5}	29	31
[28]	E.1.13	BS436043A-3	20	5.00×10^{-5}	278	337
[28]	E.1.13	BS436043A-3	100	5.00×10^{-5}	272	341
[28]	E.1.13	BS436043A-3	150	5.00×10^{-5}	272	389
[28]	E.1.13	BS436043A-3	200	5.00×10^{-5}	270	383
[28]	E.1.13	BS436043A-3	250	5.00×10^{-5}	244	381
[28]	E.1.13	BS436043A-3	300	5.00×10^{-5}	212	354
[28]	E.1.13	BS436043A-3	400	5.00×10^{-5}	205	328
[28]	E.1.13	BS436043A-3	500	5.00×10^{-5}	187	260
[28]	E.1.13	BS436043A-3	600	5.00×10^{-5}	127	140
[28]	E.1.13	BS436043A-3	700	5.00×10^{-5}	57	57
[28]	E.1.13	BS436043A-3	800	5.00×10^{-5}	37	41
[29]	E.1.15	CSA G40.21 350AT	20	1.48×10^{-5}	374	NA
[29]	E.1.15	CSA G40.21 350AT	100	1.48×10^{-5}	346	NA

NA: not available; data not calculated

$S_y^{002} = S_y(0.002 \text{ offset})$: 0.2 % offset yield strength

$S_y^{02} = S_y(0.02 \text{ elong})$: stress measured at 2 % total strain

continued on next page

Table 10: Data for structural steels from literature sources.

Continued from the previous page

Reference	Section	Steel	T	$\dot{\epsilon}$	S_y^{002}	S_y^{02}
			°C	1/s	MPa	MPa
[29]	E.1.15	CSA G40.21 350AT	150	1.48×10^{-5}	316	NA
[29]	E.1.15	CSA G40.21 350AT	200	1.48×10^{-5}	280	NA
[29]	E.1.15	CSA G40.21 350AT	250	1.48×10^{-5}	283	NA
[29]	E.1.15	CSA G40.21 350AT	300	1.48×10^{-5}	248	NA
[29]	E.1.15	CSA G40.21 350AT	20	7.41×10^{-5}	377	NA
[29]	E.1.15	CSA G40.21 350AT	100	7.41×10^{-5}	351	NA
[29]	E.1.15	CSA G40.21 350AT	150	7.41×10^{-5}	326	NA
[29]	E.1.15	CSA G40.21 350AT	200	7.41×10^{-5}	308	NA
[29]	E.1.15	CSA G40.21 350AT	250	7.41×10^{-5}	315	NA
[29]	E.1.15	CSA G40.21 350AT	300	7.41×10^{-5}	293	NA
[29]	E.1.15	CSA G40.21 350AT	20	3.71×10^{-4}	380	NA
[29]	E.1.15	CSA G40.21 350AT	100	3.71×10^{-4}	354	NA
[29]	E.1.15	CSA G40.21 350AT	150	3.71×10^{-4}	337	NA
[29]	E.1.15	CSA G40.21 350AT	200	3.71×10^{-4}	307	NA
[29]	E.1.15	CSA G40.21 350AT	250	3.71×10^{-4}	328	NA
[29]	E.1.15	CSA G40.21 350AT	300	3.71×10^{-4}	306	NA
[29]	E.1.15	CSA G40.21 350AT	20	1.48×10^{-3}	389	NA
[29]	E.1.15	CSA G40.21 350AT	100	1.48×10^{-3}	362	NA
[29]	E.1.15	CSA G40.21 350AT	150	1.48×10^{-3}	341	NA
[29]	E.1.15	CSA G40.21 350AT	200	1.48×10^{-3}	316	NA
[29]	E.1.15	CSA G40.21 350AT	250	1.48×10^{-3}	323	NA
[29]	E.1.15	CSA G40.21 350AT	300	1.48×10^{-3}	300	NA
[29]	E.1.15	CSA G40.21 350WT	20	1.48×10^{-5}	218	NA
[29]	E.1.15	CSA G40.21 350WT	100	1.48×10^{-5}	215	NA
[29]	E.1.15	CSA G40.21 350WT	150	1.48×10^{-5}	190	NA
[29]	E.1.15	CSA G40.21 350WT	200	1.48×10^{-5}	203	NA
[29]	E.1.15	CSA G40.21 350WT	250	1.48×10^{-5}	209	NA
[29]	E.1.15	CSA G40.21 350WT	300	1.48×10^{-5}	213	NA
[29]	E.1.15	CSA G40.21 350WT	350	1.48×10^{-5}	190	NA
[29]	E.1.15	CSA G40.21 350WT	20	7.41×10^{-5}	266	NA

NA: not available; data not calculated

$S_y^{002} = S_y(0.002 \text{ offset})$: 0.2 % offset yield strength

$S_y^{02} = S_y(0.02 \text{ elong})$: stress measured at 2 % total strain

continued on next page

Table 10: Data for structural steels from literature sources.

Continued from the previous page

Reference	Section	Steel	T	$\dot{\epsilon}$	S_y^{002}	S_y^{02}
			°C	1/s	MPa	MPa
[29]	E.1.15	CSA G40.21 350WT	100	7.41×10^{-5}	243	NA
[29]	E.1.15	CSA G40.21 350WT	150	7.41×10^{-5}	236	NA
[29]	E.1.15	CSA G40.21 350WT	200	7.41×10^{-5}	234	NA
[29]	E.1.15	CSA G40.21 350WT	250	7.41×10^{-5}	237	NA
[29]	E.1.15	CSA G40.21 350WT	300	7.41×10^{-5}	233	NA
[29]	E.1.15	CSA G40.21 350WT	350	7.41×10^{-5}	218	NA
[29]	E.1.15	CSA G40.21 350WT	20	3.71×10^{-4}	269	NA
[29]	E.1.15	CSA G40.21 350WT	100	3.71×10^{-4}	258	NA
[29]	E.1.15	CSA G40.21 350WT	150	3.71×10^{-4}	241	NA
[29]	E.1.15	CSA G40.21 350WT	200	3.71×10^{-4}	232	NA
[29]	E.1.15	CSA G40.21 350WT	250	3.71×10^{-4}	241	NA
[29]	E.1.15	CSA G40.21 350WT	300	3.71×10^{-4}	246	NA
[29]	E.1.15	CSA G40.21 350WT	350	3.71×10^{-4}	237	NA
[29]	E.1.15	CSA G40.21 350WT	20	1.48×10^{-3}	273	NA
[29]	E.1.15	CSA G40.21 350WT	100	1.48×10^{-3}	260	NA
[29]	E.1.15	CSA G40.21 350WT	150	1.48×10^{-3}	242	NA
[29]	E.1.15	CSA G40.21 350WT	200	1.48×10^{-3}	233	NA
[29]	E.1.15	CSA G40.21 350WT	250	1.48×10^{-3}	250	NA
[29]	E.1.15	CSA G40.21 350WT	300	1.48×10^{-3}	239	NA
[29]	E.1.15	CSA G40.21 350WT	350	1.48×10^{-3}	237	NA
[30]	E.1.17	S355	20	5.00×10^{-5}	385	397
[30]	E.1.17	S355	400	5.00×10^{-5}	NA	408
[30]	E.1.17	S355	500	5.00×10^{-5}	NA	340
[30]	E.1.17	S355	600	5.00×10^{-5}	NA	212
[30]	E.1.17	S355	700	5.00×10^{-5}	85	101
[30]	E.1.17	S355	750	5.00×10^{-5}	54	72
[30]	E.1.17	S355	800	5.00×10^{-5}	48	62
[30]	E.1.17	S355	900	5.00×10^{-5}	37	48
[31]	E.1.18	AS3679.1	20	3.33×10^{-5}	347	374
[31]	E.1.18	AS3679.1	20	3.33×10^{-5}	350	371

NA: not available; data not calculated

$S_y^{002} = S_y(0.002 \text{ offset})$: 0.2 % offset yield strength

$S_y^{02} = S_y(0.02 \text{ elong})$: stress measured at 2 % total strain

continued on next page

Table 10: Data for structural steels from literature sources.

Continued from the previous page

Reference	Section	Steel	T	$\dot{\epsilon}$	S_y^{002}	S_y^{02}
			°C	1/s	MPa	MPa
[31]	E.1.18	AS3679.1	300	3.33×10^{-5}	253	376
[31]	E.1.18	AS3679.1	300	3.33×10^{-5}	271	362
[31]	E.1.18	AS3679.1	400	3.33×10^{-5}	221	329
[31]	E.1.18	AS3679.1	400	3.33×10^{-5}	288	365
[31]	E.1.18	AS3679.1	500	3.33×10^{-5}	214	268
[31]	E.1.18	AS3679.1	500	3.33×10^{-5}	221	273
[31]	E.1.18	AS3679.1	600	3.33×10^{-5}	132	136
[31]	E.1.18	AS3679.1	600	3.33×10^{-5}	132	138
[31]	E.1.18	AS3679.1	700	3.33×10^{-5}	65	61
[31]	E.1.18	AS3679.1	700	3.33×10^{-5}	66	59
[31]	E.1.18	AS3679.1	800	3.33×10^{-5}	39	41
[31]	E.1.18	AS3679.1	20	8.00×10^{-4}	359	381
[31]	E.1.18	AS3679.1	20	8.00×10^{-4}	365	353
[31]	E.1.18	AS3679.1	400	8.00×10^{-4}	253	365
[31]	E.1.18	AS3679.1	500	8.00×10^{-4}	231	306
[31]	E.1.18	AS3679.1	600	8.00×10^{-4}	174	188
[31]	E.1.18	AS3679.1	700	8.00×10^{-4}	98	95
[31]	E.1.18	AS3679.1	800	8.00×10^{-4}	57	64
[32]	E.1.19	SM400A	20	5.00×10^{-5}	252	255
[32]	E.1.19	SM400A	100	5.00×10^{-5}	231	224
[32]	E.1.19	SM400A	200	5.00×10^{-5}	224	231
[32]	E.1.19	SM400A	300	5.00×10^{-5}	159	224
[32]	E.1.19	SM400A	400	5.00×10^{-5}	147	216
[32]	E.1.19	SM400A	500	5.00×10^{-5}	134	179
[32]	E.1.19	SM400A	550	5.00×10^{-5}	107	141
[32]	E.1.19	SM400A	600	5.00×10^{-5}	84	103
[32]	E.1.19	SM400A	650	5.00×10^{-5}	66	79
[32]	E.1.19	SM400A	700	5.00×10^{-5}	45	52
[32]	E.1.19	SM400A	750	5.00×10^{-5}	38	38
[32]	E.1.19	SM400A	800	5.00×10^{-5}	28	31

NA: not available; data not calculated

$S_y^{002} = S_y(0.002 \text{ offset}) : 0.2\ \%$ offset yield strength

$S_y^{02} = S_y(0.02 \text{ elong}) :$ stress measured at 2 % total strain

continued on next page

Table 10: Data for structural steels from literature sources.

Continued from the previous page

Reference	Section	Steel	T	$\dot{\epsilon}$	S_y^{002}	S_y^{02}
			°C	1/s	MPa	MPa
[33]	E.1.20	AS A186:250	27	1.67×10^{-7}	232	299
[33]	E.1.20	AS A186:250	27	8.33×10^{-7}	236	303
[33]	E.1.20	AS A186:250	100	8.33×10^{-7}	NA	NA
[33]	E.1.20	AS A186:250	200	8.33×10^{-7}	NA	NA
[33]	E.1.20	AS A186:250	300	8.33×10^{-7}	214	354
[33]	E.1.20	AS A186:250	350	8.33×10^{-7}	197	325
[33]	E.1.20	AS A186:250	400	8.33×10^{-7}	183	294
[33]	E.1.20	AS A186:250	450	8.33×10^{-7}	177	265
[33]	E.1.20	AS A186:250	500	8.33×10^{-7}	119	219
[33]	E.1.20	AS A186:250	550	8.33×10^{-7}	99	133
[33]	E.1.20	AS A186:250	600	8.33×10^{-7}	80	76
[33]	E.1.20	AS A186:250	650	8.33×10^{-7}	45	46
[33]	E.1.20	AS A186:250	350	8.33×10^{-6}	190	NA
[33]	E.1.20	AS A186:250	400	8.33×10^{-6}	199	NA
[33]	E.1.20	AS A186:250	450	8.33×10^{-6}	185	NA
[33]	E.1.20	AS A186:250	500	8.33×10^{-6}	170	NA
[33]	E.1.20	AS A186:250	550	8.33×10^{-6}	128	NA
[33]	E.1.20	AS A186:250	600	8.33×10^{-6}	91	NA
[33]	E.1.20	AS A186:250	650	8.33×10^{-6}	63	NA
[33]	E.1.20	AS A186:250	27	3.33×10^{-5}	250	319
[33]	E.1.20	AS A186:250	100	3.33×10^{-5}	232	327
[33]	E.1.20	AS A186:250	200	3.33×10^{-5}	214	375
[33]	E.1.20	AS A186:250	300	3.33×10^{-5}	202	345
[33]	E.1.20	AS A186:250	350	3.33×10^{-5}	201	339
[33]	E.1.20	AS A186:250	400	3.33×10^{-5}	202	329
[33]	E.1.20	AS A186:250	450	3.33×10^{-5}	194	297
[33]	E.1.20	AS A186:250	500	3.33×10^{-5}	181	254
[33]	E.1.20	AS A186:250	550	3.33×10^{-5}	145	181
[33]	E.1.20	AS A186:250	600	3.33×10^{-5}	108	121
[33]	E.1.20	AS A186:250	650	3.33×10^{-5}	77	82

NA: not available; data not calculated

$S_y^{002} = S_y(0.002 \text{ offset})$: 0.2 % offset yield strength

$S_y^{02} = S_y(0.02 \text{ elong})$: stress measured at 2 % total strain

continued on next page

Table 10: Data for structural steels from literature sources.

Continued from the previous page

Reference	Section	Steel	T	$\dot{\epsilon}$	S_y^{002}	S_y^{02}
			°C	1/s	MPa	MPa
[33]	E.1.20	AS A186:250	27	3.33×10^{-4}	263	334
[33]	E.1.20	AS A186:250	27	3.33×10^{-3}	NA	327
[33]	E.1.20	AS A186:250	350	3.33×10^{-3}	NA	340
[33]	E.1.20	AS A186:250	500	3.33×10^{-3}	NA	278
[33]	E.1.20	AS A186:250	650	3.33×10^{-3}	NA	142
[34]	E.1.22	Cor-Ten A 10	27	8.33×10^{-5}	368	NA
[34]	E.1.22	Cor-Ten A 10	93	8.33×10^{-5}	346	NA
[34]	E.1.22	Cor-Ten A 10	204	8.33×10^{-5}	323	NA
[34]	E.1.22	Cor-Ten A 10	316	8.33×10^{-5}	281	NA
[34]	E.1.22	Cor-Ten A 10	427	8.33×10^{-5}	258	NA
[34]	E.1.22	Cor-Ten A 10	538	8.33×10^{-5}	239	NA
[34]	E.1.22	Cor-Ten A 10	649	8.33×10^{-5}	139	NA
[34]	E.1.22	Cor-Ten A 6	27	8.33×10^{-5}	354	NA
[34]	E.1.22	Cor-Ten A 6	93	8.33×10^{-5}	327	NA
[34]	E.1.22	Cor-Ten A 6	204	8.33×10^{-5}	316	NA
[34]	E.1.22	Cor-Ten A 6	316	8.33×10^{-5}	277	NA
[34]	E.1.22	Cor-Ten A 6	427	8.33×10^{-5}	243	NA
[34]	E.1.22	Cor-Ten A 6	538	8.33×10^{-5}	224	NA
[34]	E.1.22	Cor-Ten A 6	649	8.33×10^{-5}	135	NA
[34]	E.1.22	Cor-Ten A 7	27	8.33×10^{-5}	396	NA
[34]	E.1.22	Cor-Ten A 7	93	8.33×10^{-5}	376	NA
[34]	E.1.22	Cor-Ten A 7	204	8.33×10^{-5}	352	NA
[34]	E.1.22	Cor-Ten A 7	316	8.33×10^{-5}	304	NA
[34]	E.1.22	Cor-Ten A 7	427	8.33×10^{-5}	291	NA
[34]	E.1.22	Cor-Ten A 7	538	8.33×10^{-5}	262	NA
[34]	E.1.22	Cor-Ten A 7	649	8.33×10^{-5}	149	NA
[34]	E.1.22	Cor-Ten A 8	27	8.33×10^{-5}	445	NA
[34]	E.1.22	Cor-Ten A 8	93	8.33×10^{-5}	414	NA
[34]	E.1.22	Cor-Ten A 8	204	8.33×10^{-5}	381	NA
[34]	E.1.22	Cor-Ten A 8	316	8.33×10^{-5}	337	NA

NA: not available; data not calculated

$S_y^{002} = S_y(0.002 \text{ offset})$: 0.2 % offset yield strength

$S_y^{02} = S_y(0.02 \text{ elong})$: stress measured at 2 % total strain

continued on next page

82

Table 10: Data for structural steels from literature sources.

Continued from the previous page

Reference	Section	Steel	T	$\dot{\epsilon}$	S_y^{002}	S_y^{02}
			°C	1/s	MPa	MPa
[34]	E.1.22	Cor-Ten A 8	427	8.33×10^{-5}	296	NA
[34]	E.1.22	Cor-Ten A 8	538	8.33×10^{-5}	241	NA
[34]	E.1.22	Cor-Ten A 8	649	8.33×10^{-5}	128	NA
[34]	E.1.22	Cor-Ten B 12	27	8.33×10^{-5}	439	NA
[34]	E.1.22	Cor-Ten B 12	93	8.33×10^{-5}	416	NA
[34]	E.1.22	Cor-Ten B 12	204	8.33×10^{-5}	374	NA
[34]	E.1.22	Cor-Ten B 12	316	8.33×10^{-5}	375	NA
[34]	E.1.22	Cor-Ten B 12	427	8.33×10^{-5}	364	NA
[34]	E.1.22	Cor-Ten B 12	538	8.33×10^{-5}	287	NA
[34]	E.1.22	Cor-Ten B 12	649	8.33×10^{-5}	167	NA
[34]	E.1.22	Cor-Ten B 13	27	8.33×10^{-5}	399	NA
[34]	E.1.22	Cor-Ten B 13	93	8.33×10^{-5}	376	NA
[34]	E.1.22	Cor-Ten B 13	204	8.33×10^{-5}	338	NA
[34]	E.1.22	Cor-Ten B 13	316	8.33×10^{-5}	332	NA
[34]	E.1.22	Cor-Ten B 13	427	8.33×10^{-5}	317	NA
[34]	E.1.22	Cor-Ten B 13	538	8.33×10^{-5}	283	NA
[34]	E.1.22	Cor-Ten B 13	649	8.33×10^{-5}	189	NA
[34]	E.1.22	Cor-Ten B 14	27	8.33×10^{-5}	415	NA
[34]	E.1.22	Cor-Ten B 14	93	8.33×10^{-5}	390	NA
[34]	E.1.22	Cor-Ten B 14	204	8.33×10^{-5}	355	NA
[34]	E.1.22	Cor-Ten B 14	316	8.33×10^{-5}	339	NA
[34]	E.1.22	Cor-Ten B 14	427	8.33×10^{-5}	323	NA
[34]	E.1.22	Cor-Ten B 14	538	8.33×10^{-5}	253	NA
[34]	E.1.22	Cor-Ten B 14	649	8.33×10^{-5}	151	NA
[34]	E.1.22	Cor-Ten B 15	27	8.33×10^{-5}	367	NA
[34]	E.1.22	Cor-Ten B 15	93	8.33×10^{-5}	350	NA
[34]	E.1.22	Cor-Ten B 15	204	8.33×10^{-5}	336	NA
[34]	E.1.22	Cor-Ten B 15	316	8.33×10^{-5}	331	NA
[34]	E.1.22	Cor-Ten B 15	427	8.33×10^{-5}	306	NA
[34]	E.1.22	Cor-Ten B 15	538	8.33×10^{-5}	243	NA

NA: not available; data not calculated

$S_y^{002} = S_y(0.002 \text{ offset})$: 0.2 % offset yield strength

$S_y^{02} = S_y(0.02 \text{ elong})$: stress measured at 2 % total strain

continued on next page

Table 10: Data for structural steels from literature sources.

Continued from the previous page

Reference	Section	Steel	T	$\dot{\epsilon}$	S_y^{002}	S_y^{02}
			°C	1/s	MPa	MPa
[34]	E.1.22	Cor-Ten B 15	649	8.33×10^{-5}	149	NA
[6]	E.1.23	AS187:WR350/1	27	1.67×10^{-7}	391	394
[6]	E.1.23	AS187:WR350/1	350	1.67×10^{-7}	292	360
[6]	E.1.23	AS187:WR350/1	350	1.67×10^{-7}	302	361
[6]	E.1.23	AS187:WR350/1	500	1.67×10^{-7}	173	183
[6]	E.1.23	AS187:WR350/1	500	1.67×10^{-7}	177	190
[6]	E.1.23	AS187:WR350/1	650	1.67×10^{-7}	64	64
[6]	E.1.23	AS187:WR350/1	650	1.67×10^{-7}	66	61
[6]	E.1.23	AS187:WR350/1	27	8.33×10^{-7}	392	394
[6]	E.1.23	AS187:WR350/1	27	8.33×10^{-7}	392	395
[6]	E.1.23	AS187:WR350/1	100	8.33×10^{-7}	375	389
[6]	E.1.23	AS187:WR350/1	200	8.33×10^{-7}	359	404
[6]	E.1.23	AS187:WR350/1	200	8.33×10^{-7}	361	405
[6]	E.1.23	AS187:WR350/1	300	8.33×10^{-7}	314	378
[6]	E.1.23	AS187:WR350/1	350	8.33×10^{-7}	304	365
[6]	E.1.23	AS187:WR350/1	400	8.33×10^{-7}	291	347
[6]	E.1.23	AS187:WR350/1	450	8.33×10^{-7}	270	317
[6]	E.1.23	AS187:WR350/1	475	8.33×10^{-7}	248	279
[6]	E.1.23	AS187:WR350/1	500	8.33×10^{-7}	203	223
[6]	E.1.23	AS187:WR350/1	550	8.33×10^{-7}	148	159
[6]	E.1.23	AS187:WR350/1	600	8.33×10^{-7}	111	114
[6]	E.1.23	AS187:WR350/1	650	8.33×10^{-7}	81	79
[6]	E.1.23	AS187:WR350/1	27	8.33×10^{-6}	407	405
[6]	E.1.23	AS187:WR350/1	350	8.33×10^{-6}	308	374
[6]	E.1.23	AS187:WR350/1	400	8.33×10^{-6}	291	354
[6]	E.1.23	AS187:WR350/1	450	8.33×10^{-6}	281	336
[6]	E.1.23	AS187:WR350/1	500	8.33×10^{-6}	246	272
[6]	E.1.23	AS187:WR350/1	550	8.33×10^{-6}	179	192
[6]	E.1.23	AS187:WR350/1	600	8.33×10^{-6}	136	143
[6]	E.1.23	AS187:WR350/1	650	8.33×10^{-6}	NA	102

NA: not available; data not calculated

$S_y^{002} = S_y(0.002 \text{ offset})$: 0.2 % offset yield strength

$S_y^{02} = S_y(0.02 \text{ elong})$: stress measured at 2 % total strain

continued on next page

84

Table 10: Data for structural steels from literature sources.

Continued from the previous page

Reference	Section	Steel	T	$\dot{\epsilon}$	S_y^{002}	S_y^{02}
			°C	1/s	MPa	MPa
[6]	E.1.23	AS187:WR350/1	27	3.33×10^{-5}	398	430
[6]	E.1.23	AS187:WR350/1	27	3.33×10^{-5}	398	401
[6]	E.1.23	AS187:WR350/1	100	3.33×10^{-5}	379	383
[6]	E.1.23	AS187:WR350/1	200	3.33×10^{-5}	370	401
[6]	E.1.23	AS187:WR350/1	300	3.33×10^{-5}	339	401
[6]	E.1.23	AS187:WR350/1	350	3.33×10^{-5}	309	371
[6]	E.1.23	AS187:WR350/1	400	3.33×10^{-5}	296	355
[6]	E.1.23	AS187:WR350/1	450	3.33×10^{-5}	283	336
[6]	E.1.23	AS187:WR350/1	450	3.33×10^{-5}	283	348
[6]	E.1.23	AS187:WR350/1	475	3.33×10^{-5}	270	319
[6]	E.1.23	AS187:WR350/1	500	3.33×10^{-5}	263	294
[6]	E.1.23	AS187:WR350/1	550	3.33×10^{-5}	204	220
[6]	E.1.23	AS187:WR350/1	550	3.33×10^{-5}	203	217
[6]	E.1.23	AS187:WR350/1	600	3.33×10^{-5}	154	159
[6]	E.1.23	AS187:WR350/1	650	3.33×10^{-5}	112	113
[6]	E.1.23	AS187:WR350/1	650	3.33×10^{-5}	114	115
[6]	E.1.23	AS187:WR350/1	27	3.33×10^{-4}	409	408
[6]	E.1.23	AS187:WR350/1	350	3.33×10^{-4}	300	364
[6]	E.1.23	AS187:WR350/1	500	3.33×10^{-4}	273	316
[6]	E.1.23	AS187:WR350/1	650	3.33×10^{-4}	139	144
[6]	E.1.23	AS187:WR350/1	27	3.33×10^{-3}	NA	416
[6]	E.1.23	AS187:WR350/1	350	3.33×10^{-3}	323	367
[6]	E.1.23	AS187:WR350/1	350	3.33×10^{-3}	298	363
[6]	E.1.23	AS187:WR350/1	500	3.33×10^{-3}	NA	NA
[6]	E.1.23	AS187:WR350/1	500	3.33×10^{-3}	301	350
[6]	E.1.23	AS187:WR350/1	500	3.33×10^{-3}	290	330
[6]	E.1.23	AS187:WR350/1	650	3.33×10^{-3}	NA	179
[6]	E.1.23	AS187:WR350/1	650	3.33×10^{-3}	170	179
[6]	E.1.23	AS A186:250 L0	27	8.33×10^{-7}	232	308
[6]	E.1.23	AS A186:250 L0	350	8.33×10^{-7}	192	310

NA: not available; data not calculated

$S_y^{002} = S_y(0.002 \text{ offset})$: 0.2 % offset yield strength

$S_y^{02} = S_y(0.02 \text{ elong})$: stress measured at 2 % total strain

continued on next page

Table 10: Data for structural steels from literature sources.

Continued from the previous page

Reference	Section	Steel	T	$\dot{\epsilon}$	S_y^{002}	S_y^{02}
			°C	1/s	MPa	MPa
[6]	E.1.23	AS A186:250 L0	400	8.33×10^{-7}	186	289
[6]	E.1.23	AS A186:250 L0	450	8.33×10^{-7}	196	276
[6]	E.1.23	AS A186:250 L0	500	8.33×10^{-7}	174	223
[6]	E.1.23	AS A186:250 L0	550	8.33×10^{-7}	120	136
[6]	E.1.23	AS A186:250 L0	600	8.33×10^{-7}	72	79
[6]	E.1.23	AS A186:250 L0	650	8.33×10^{-7}	43	46
[6]	E.1.23	AS A186:250 L0	27	3.33×10^{-5}	226	303
[6]	E.1.23	AS A186:250 L0	350	3.33×10^{-5}	221	355
[6]	E.1.23	AS A186:250 L0	400	3.33×10^{-5}	198	319
[6]	E.1.23	AS A186:250 L0	450	3.33×10^{-5}	209	311
[6]	E.1.23	AS A186:250 L0	500	3.33×10^{-5}	176	248
[6]	E.1.23	AS A186:250 L0	550	3.33×10^{-5}	NA	196
[6]	E.1.23	AS A186:250 L0	550	3.33×10^{-5}	157	216
[6]	E.1.23	AS A186:250 L0	600	3.33×10^{-5}	115	133
[6]	E.1.23	AS A186:250 L0	650	3.33×10^{-5}	80	87
[6]	E.1.23	AS A186:400 L15	27	1.67×10^{-7}	414	418
[6]	E.1.23	AS A186:400 L15	27	8.33×10^{-7}	416	414
[6]	E.1.23	AS A186:400 L15	300	8.33×10^{-7}	286	432
[6]	E.1.23	AS A186:400 L15	350	8.33×10^{-7}	283	414
[6]	E.1.23	AS A186:400 L15	400	8.33×10^{-7}	260	368
[6]	E.1.23	AS A186:400 L15	450	8.33×10^{-7}	236	309
[6]	E.1.23	AS A186:400 L15	500	8.33×10^{-7}	191	223
[6]	E.1.23	AS A186:400 L15	550	8.33×10^{-7}	126	135
[6]	E.1.23	AS A186:400 L15	600	8.33×10^{-7}	77	78
[6]	E.1.23	AS A186:400 L15	650	8.33×10^{-7}	45	44
[6]	E.1.23	AS A186:400 L15	27	3.33×10^{-5}	430	430
[6]	E.1.23	AS A186:400 L15	100	3.33×10^{-5}	400	399
[6]	E.1.23	AS A186:400 L15	200	3.33×10^{-5}	371	399
[6]	E.1.23	AS A186:400 L15	300	3.33×10^{-5}	292	462
[6]	E.1.23	AS A186:400 L15	350	3.33×10^{-5}	292	427

NA: not available; data not calculated

$S_y^{002} = S_y(0.002 \text{ offset})$: 0.2 % offset yield strength

$S_y^{02} = S_y(0.02 \text{ elong})$: stress measured at 2 % total strain

continued on next page

86

Table 10: Data for structural steels from literature sources.

Continued from the previous page

Reference	Section	Steel	T	$\dot{\epsilon}$	S_y^{002}	S_y^{02}
			°C	1/s	MPa	MPa
[6]	E.1.23	AS A186:400 L15	400	3.33×10^{-5}	280	423
[6]	E.1.23	AS A186:400 L15	450	3.33×10^{-5}	254	392
[6]	E.1.23	AS A186:400 L15	500	3.33×10^{-5}	233	352
[6]	E.1.23	AS A186:400 L15	550	3.33×10^{-5}	192	291
[6]	E.1.23	AS A186:400 L15	600	3.33×10^{-5}	137	219
[6]	E.1.23	AS A186:400 L15	650	3.33×10^{-5}	92	143
[6]	E.1.23	AS A186:400 L15	27	3.33×10^{-4}	444	93
[6]	E.1.23	AS A186:400 L15	27	3.33×10^{-3}	450	445
[6]	E.1.23	AS A186:400 L15	350	3.33×10^{-3}	NA	447
[6]	E.1.23	AS A186:400 L15	500	3.33×10^{-3}	NA	421
[6]	E.1.23	AS A186:400 L15	650	3.33×10^{-3}	NA	341
[35]	E.1.24	USS A36	27	8.33×10^{-5}	248	NA
[35]	E.1.24	USS A36	149	8.33×10^{-5}	208	NA
[35]	E.1.24	USS A36	260	8.33×10^{-5}	192	NA
[35]	E.1.24	USS A36	371	8.33×10^{-5}	175	NA
[35]	E.1.24	USS A36	482	8.33×10^{-5}	148	NA
[35]	E.1.24	USS A36	593	8.33×10^{-5}	112	NA
[35]	E.1.24	USS A36	704	8.33×10^{-5}	53	NA
[35]	E.1.24	USS T1	27	8.30×10^{-5}	803	NA
[35]	E.1.24	USS T1	93	8.30×10^{-5}	767	NA
[35]	E.1.24	USS T1	204	8.30×10^{-5}	703	NA
[35]	E.1.24	USS T1	316	8.30×10^{-5}	678	NA
[35]	E.1.24	USS T1	427	8.30×10^{-5}	618	NA
[35]	E.1.24	USS T1	538	8.30×10^{-5}	519	NA
[35]	E.1.24	USS T1	649	8.30×10^{-5}	241	NA
[35]	E.1.24	USS T1	760	8.30×10^{-5}	63	NA
[35]	E.1.24	USS T1	871	8.30×10^{-5}	46	NA
[35]	E.1.24	USS T1	1038	8.30×10^{-5}	19	NA
[62]	E.1.16	mild steel	20	9.50×10^{-7}	190	236
[62]	E.1.16	mild steel	200	3.40×10^{-5}	177	276

NA: not available; data not calculated

$S_y^{002} = S_y(0.002 \text{ offset})$: 0.2 % offset yield strength

$S_y^{02} = S_y(0.02 \text{ elong})$: stress measured at 2 % total strain

continued on next page

Table 10: Data for structural steels from literature sources.

Continued from the previous page

Reference	Section	Steel	T	$\dot{\epsilon}$	S_y^{002}	S_y^{02}
			°C	1/s	MPa	MPa
[62]	E.1.16	mild steel	600	3.40×10^{-5}	53	75
[62]	E.1.16	mild steel	20	8.50×10^{-4}	214	250
[62]	E.1.16	mild steel	200	8.50×10^{-4}	181	307
[62]	E.1.16	mild steel	400	8.50×10^{-4}	211	274
[62]	E.1.16	mild steel	200	2.00×10^{-2}	188	314
[62]	E.1.16	mild steel	600	2.00×10^{-2}	112	162
[62]	E.1.16	mild steel	20	5.00×10^{-1}	259	290
[62]	E.1.16	mild steel	200	5.00×10^{-1}	228	276
[62]	E.1.16	mild steel	400	5.00×10^{-1}	233	284
[62]	E.1.16	mild steel	20	1.00×10^{2}	452	NA
[62]	E.1.16	mild steel	400	1.87×10^{2}	274	329
[62]	E.1.16	mild steel	200	2.00×10^{2}	392	416
[62]	E.1.16	mild steel	600	2.00×10^{2}	303	349
[62]	E.1.16	mild steel	20	3.02×10^{2}	506	508
[62]	E.1.16	mild steel	400	6.00×10^{2}	281	339
[40]	E.1.10	A992	20	1.67×10^{-4}	429	425
[40]	E.1.10	A992	100	1.67×10^{-4}	409	408
[40]	E.1.10	A992	200	1.67×10^{-4}	406	425
[40]	E.1.10	A992	300	1.67×10^{-4}	315	418
[40]	E.1.10	A992	400	1.67×10^{-4}	298	395
[40]	E.1.10	A992	500	1.67×10^{-4}	254	316
[40]	E.1.10	A992	600	1.67×10^{-4}	171	193
[40]	E.1.10	A992	700	1.67×10^{-4}	88	89
[40]	E.1.10	A992	800	1.67×10^{-4}	37	39
[40]	E.1.10	A992	900	1.67×10^{-4}	27	30
[40]	E.1.10	A992	20	1.67×10^{-3}	436	438
[40]	E.1.10	A992	100	1.67×10^{-3}	406	406
[40]	E.1.10	A992	200	1.67×10^{-3}	401	419
[40]	E.1.10	A992	300	1.67×10^{-3}	332	431
[40]	E.1.10	A992	400	1.67×10^{-3}	297	383

NA: not available; data not calculated

$S_y^{002} = S_y(0.002 \text{ offset})$: 0.2 % offset yield strength

$S_y^{02} = S_y(0.02 \text{ elong})$: stress measured at 2 % total strain

continued on next page

Table 10: Data for structural steels from literature sources.

Continued from the previous page

Reference	Section	Steel	T	$\dot{\epsilon}$	S_y^{002}	S_y^{02}
			°C	1/s	MPa	MPa
[40]	E.1.10	A992	500	1.67×10^{-3}	255	322
[40]	E.1.10	A992	600	1.67×10^{-3}	191	216
[40]	E.1.10	A992	700	1.67×10^{-3}	112	116
[40]	E.1.10	A992	800	1.67×10^{-3}	54	64
[40]	E.1.10	A992	900	1.67×10^{-3}	45	48

NA: not available

S_y(0.002 offset) : 0.2 % offset yield strength

S_y(0.02 elong) : stress measured at 2 % total strain

E.3   Retained strength

Figure 2 overlays all the data on a single plot. Figure 30 replots the Figure 2 data, identified by source [6,12,18,23–35]. Within each panel, different symbols denote individual steels from the same literature source.

E.4   Elastic Modulus

E.4.1   Recommended value

The NIST World Trade Center collapse investigation [12] (Chapter 2) determined the elastic modulus for three plate steels using a thermo-mechanical analyzer operating in three-point bending at 1 hz. Specimens were nominally $(50 \times 10 \times 1)$ mm. The temperature range was $(-140 < T < 600)$ °C. Table 11 summarizes the information on the steels. Figure 31 plots the data for the five determinations of the modulus.

The data were fit with a four-term polynomial:

$$E = E_0 + e_1 T + e_2 T^2 + e_3 T^3 \tag{17}$$

where

$E_0 = 206$ GPa is the elastic modulus at $T = 0$ °C.
$e_1 = -4.326 \times 10^{-2}$ GPa/°C^{-1}
$e_2 = -3.502 \times 10^{-5}$ GPa/°C^{-2}
$e_3 = -6.592 \times 10^{-8}$ GPa/°C^{-3}

The intercept, E_0, was not fit, but instead set to the average value of the $T = 0$ °C elastic modulus of the individual determinations. Figure 31 also overplots the fit of Eq. (6) to the data for each steel. In all cases the maximum deviation of the measured modulus and the fit is less than 4 GPa.

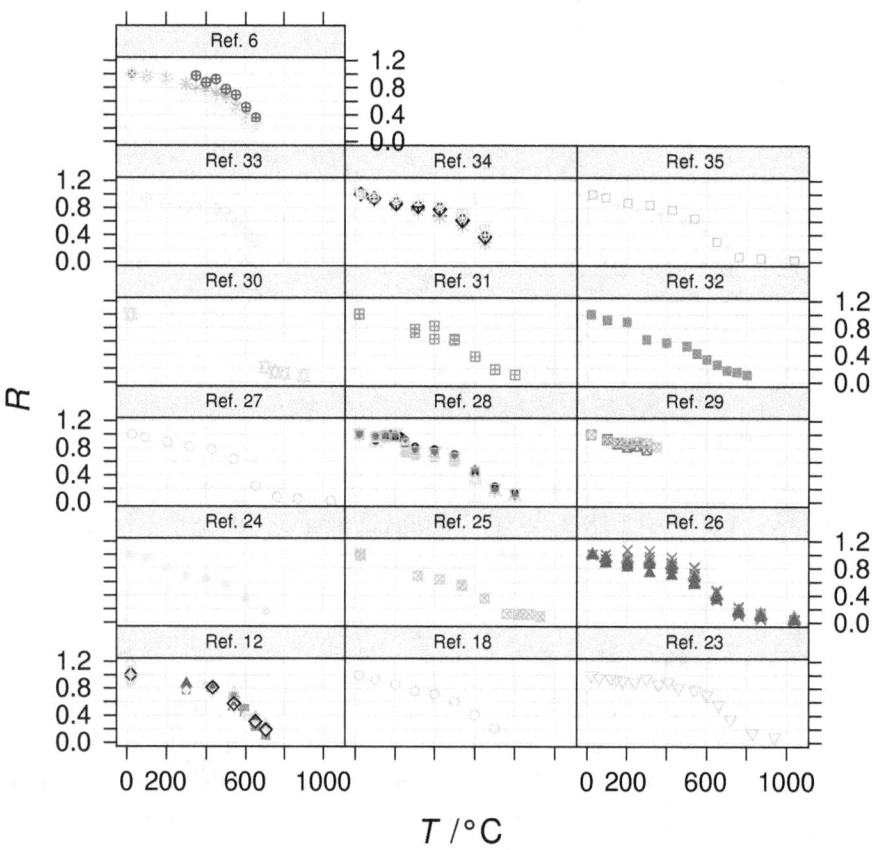

Figure 30: Normalized, retained yield strength, R, as a function of temperature, from Figure 2. Strips in each panel identify the source from the reference list. Sources: [6, 12, 18, 23–35].

Table 11: Data for the three steels

Designation	No. of runs	F_y
		ksi
C68-C3T1-EP-2	2	50
N9-C2B-FR-1	1	55
S9-C3T-FL-1	2	50

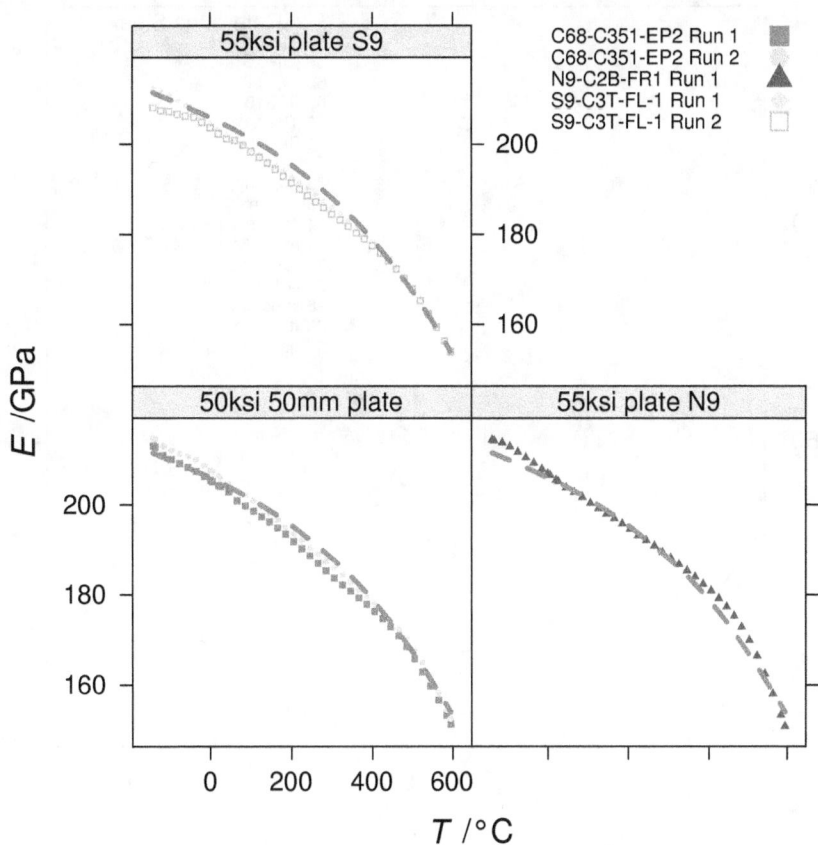

Figure 31: Temperature dependence of the elastic modulus of structural steel deter-mined from the NIST WTC collapse investigation. Dashed line is the fit of Eq.(6).

E.4.2 Other determinations and representations of elastic modulus

Investigators have determined the temperature dependence of the elastic modulus of structural steel many times in the past century, using a variety of techniques. Lie [15] summarizes some historical measurements of elastic modulus; two of these [66,67] date back into the early 20th century and are not included in the plots of this section. Figure 32 plots the change in elastic modulus with temperature for various structural steels and pure iron determined by various techniques. Figure 33 breaks out the data of Figure 32 into individual plots by reference number. Table 12

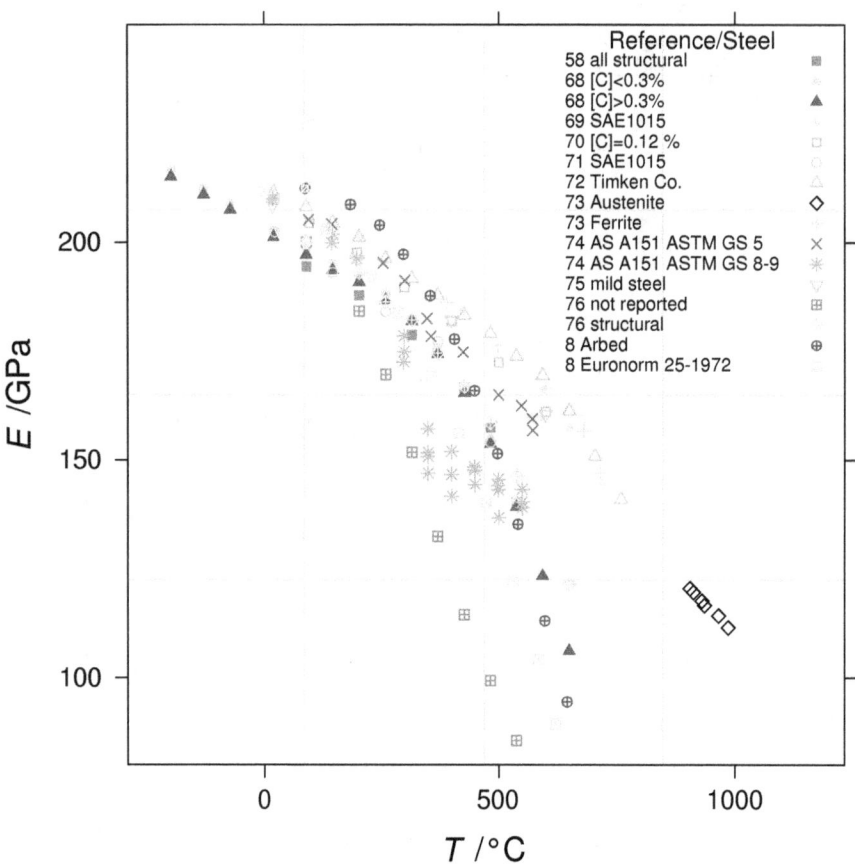

Figure 32: Literature determinations of the temperature dependence of the elastic modulus of structural steel.

summarizes the literature sources for data used in Figure 32; Section E.4.3 describes the steels and determinations in greater detail. Finally, Table 13 at the end of this section tabulates the values of the modulus used in the individual figures.

The modulus data naturally fall into two groups determined by the method used to measure the modulus. Figure 34 replots the data of Figure 32 but breaks it into three measurement method groups: dynamic, static, and not reported. Values measured by various dynamic methods [69, 70, 72, 73, 75], generally at around 400 hz, are the largest, and drop off slowest with increasing temperature. Values measured in tensile tests are labelled "static." The types labeled "not reported"

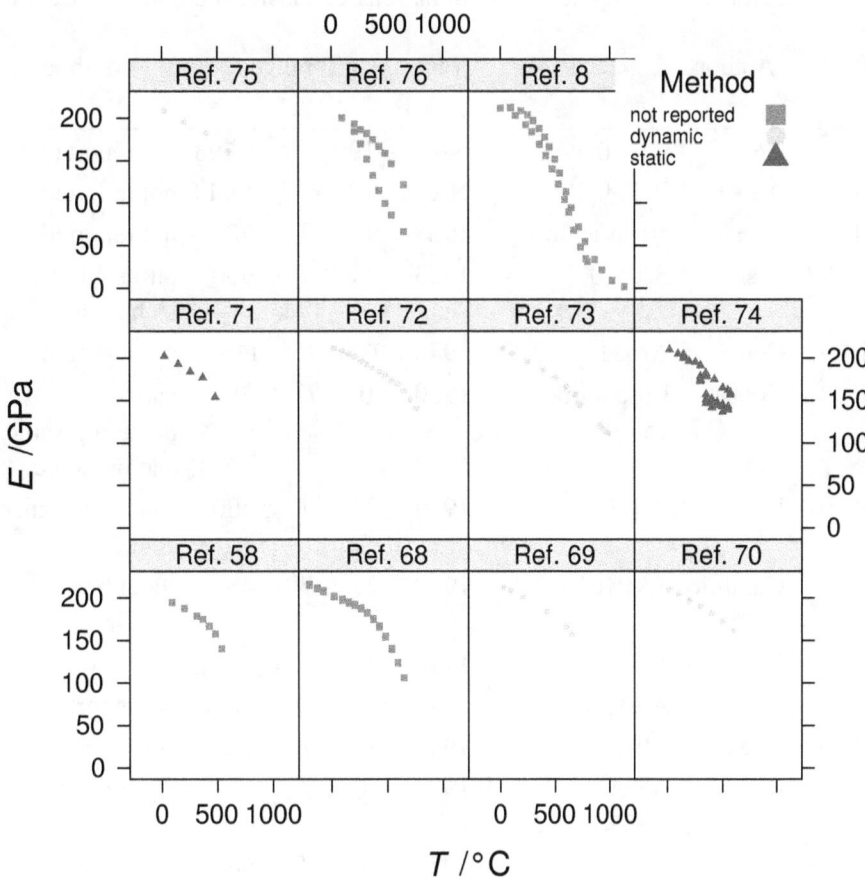

Figure 33: Literature determinations of the temperature dependence of the elastic modulus of structural steel. Colors and symbols denote the measurement method, and the strip in each plot notes the reference number.

[8, 58, 76] are also probably derived from quasi-static tests tensile tests, where the modulus is determined from the stress-strain curve. However, the sources do not identify the method. The static tests can contain significant anelastic, or time-dependent elastic effects, for example in the data of Ritter [74].

E.4.3   Notes on literature reports of recommended values for elastic modulus

Figure 35 plots the recommended modulus from four organizations concerned with the performance of steel in fire: Eq. (6) used in the WTC investigation [12], the rec-

Table 12: Summary of literature determinations of elastic modulus of structural steel.

Ref.	Pg.	Author	Steel	Year	T range °C	Method
[68]	93	ASME	[C]<0.3%	NA	$-198 < T < 593$	not reported
[68]	93	ASME	[C]>0.3%	NA	$-198 < T < 649$	not reported
[58]	96	USS	all structural	1968	$93 < T < 538$	not reported
[69]	97	Clark	SAE1015	1953	$23 < T < 650$	ultrasonic pulse technique
[8]	97	Cooke	Arbed	1986	$91 < T < 1135$	not reported
[8]	97	Cooke	Euronorm 25-1972	1980	$0 < T < 793$	various including static and dynamic
[70]	97	Date	[C]=0.12 %	1969	$20 < T < 600$	sonic resonance at 400 hz
[71]	97	Garafolo	SAE1015	1952	$24 < T < 482$	static beam bending
[72]	97	Garafolo	Timken Co.	1948	$22 < T < 760$	NA
[73]	97	Koester	Austenite	1948	$905 < T < 985$	not reported
[73]	97	Koester	Ferrite	1948	$20 < T < 720$	not reported
[74]	97	Ritter	AS A151 ASTM GS 5	1970	$20 < T < 572$	tensile test
[74]	97	Ritter	AS A151 ASTM GS 8-9	1970	$20 < T < 550$	tensile test
[75]	97	Roberts	mild steel	1947	$20 < T < 600$	sonic resonance at 400 hz
[76]	98	Uddin	not reported	NA	$93 < T < 649$	not reported
[76]	98	Uddin	structural	1961	$93 < T < 649$	not reported

ommended value of the Eurocode 3 [1], the recommended value of the American Society of Civil Engineers (ASCE) [77], and the recommended value of the European Convention on Constructional Steelwork (ECCS) [9]. References 58,68. The experimental data of Figure 32 are plotted as symbols. None of the original references, summarized below, document the rationale for choosing the specific form of the behavior.

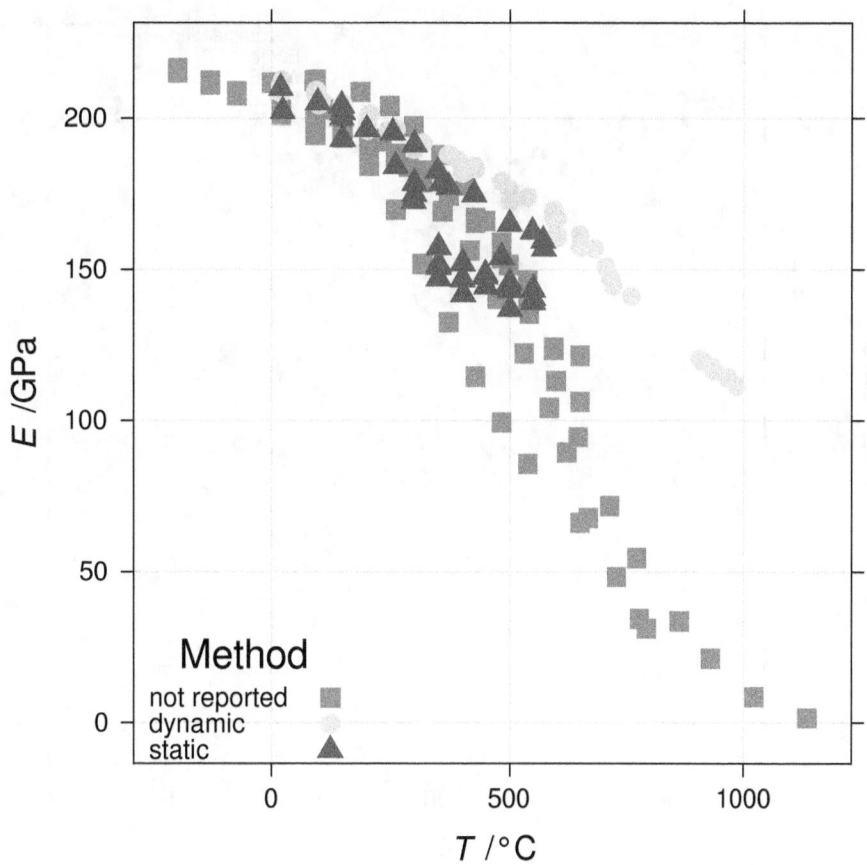

Figure 34: Temperature dependence of the elastic modulus of structural steel identified by the measurement method.

ASME The American Society of Mechanical Engineers (ASME) Boiler and Pressure Vessel Code [68] provides typical values for many physical properties, including elastic modulus. The Code describes the data as tending "to be closer to average values," but does not describe the source or the measurement technique. The data are segregated in to steels with carbon content above and below 0.3 % by weight in the temperature range $(-198 < T < 593)$ °C. However, the difference between the two is 1.4 GPa at all temperatures.

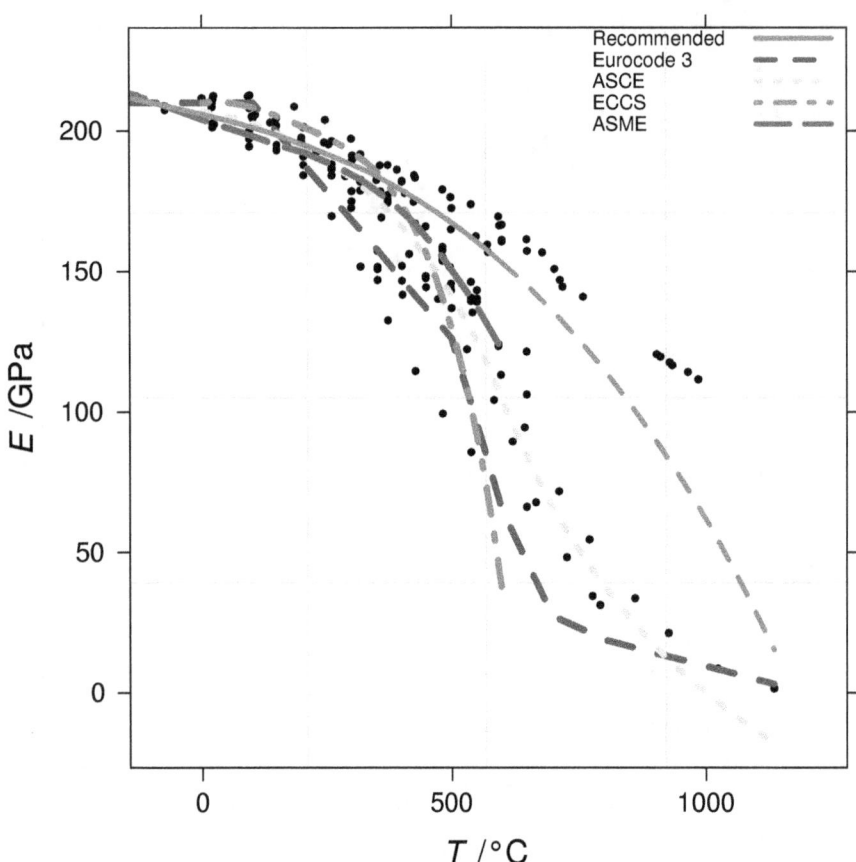

Figure 35: Values of the recommended elastic modulus from various organizations: Eurocode 3 [1], ASCE [77], and ECCS [9].

ASCE The American Society of Civil Engineers (ASCE) Structural Fire Protection Manual [77], Appendix, page 225 and Figure 2.7, see also Kodur [78], recommends an an expression

$$
E = \begin{cases} E_0 \left(1 + \dfrac{T}{e_1 \log_e\left(\dfrac{T}{e_2}\right)} \right) & T \leq 600\,^\circ\mathrm{C} \\[4ex] E_0 \left(\dfrac{e_3 - e_4 T}{T - e_5} \right) & T > 600\,^\circ\mathrm{C} \end{cases}
\tag{18}
$$

where $e_1 = 2000\,°C$, $e_2 = 1100\,°C$, $e_3 = 690$, $e_4 = 0.69\,°C^{-1}$, $e_5 = 53.5\,°C$. The value of the function goes to zero at $1000\,°C$, but then continues, unphysically, into negative modulus for higher temperatures. The appendix of the manual does not discuss either the origin or limitations of this expression. The text contains a reference number, but that number does not appear in the corresponding reference list. The values predicted by Eq. (18) are drop off more quickly than the smoothed curve that Lie, the editor of the Structural Fire Protection Manual, presented in his 1974 paper [15].

Eurocode 3   Section F explains the functional dependence of the elastic modulus of the Eurocode 3 model [1]. The origins of the parameters are not documented.

ECCS   The European Convention for Constructional Steelwork, ECCS, published a stress-strain model in 1983 [9] that includes an elastic term.

$$E_{\text{ECCS}} = E_0 \left(1 + e_1 T + e_2 T^2 + e_3 T^3 + e_4 T^4\right) \tag{19}$$

where

$e_1 = +15.9 \times 10^{-5}\,°C^{-1}$
$e_2 = -34.5 \times 10^{-7}\,°C^{-2}$
$e_3 = +11.8 \times 10^{-9}\,°C^{-3}$
$e_4 = -17.2 \times 10^{-12}\,°C^{-4}$
E_0 is the elastic modulus at $T = 0\,°C$.

Figure r-3 in the ECCS report [9] shows that $E_0 = 210$ GPa. The function is explicitly undefined for $T > 600\,°C$. The origin of the data used to model the material behavior is undocumented, other than that it was "based on data obtained in the Netherlands and elsewhere." Appendix G summarizes the stress-strain behavior the ECCS model.

E.4.4   Notes on literature reports of elastic modulus measurements

The following paragraphs summarize the methods and limitations of the data for the steels that appear in Figure 32.

United States Steel (USS)   The United States Steel Corp. "Steel Design Manual" [58] recommends a value of $E = 200$ GPa ($E = 29 \times 10^6$ psi) for the room-temperature elastic modulus for all steels (p. 8). Figure 1.7 in the Steel Design Manual graphically represents the behavior of the elastic modulus up to $T = 538\,°C$. It does not state how the modulus was measured. Because Garofalo [71, 72] (see below) worked for United States Steel, it is possible that the values in the Steel Design Manual are the values he reported earlier, and that any differences represent digitization errors from the graphical plots.

Clark   Clark [69] reprinted elastic modulus data for SAE 1015 steel measured using an ultrasonic pulse technique, from Frederick's [79] thesis and conference report in the temperature range $(23 < T < 650)$ °C.

Cooke   Cooke [8] presented a smooth curve of the variation of elastic modulus in the temperature range $(0 < T < 793)$ °C from an unpublished 1980 survey made by Stirland of British Steel Corp. using "many dynamically and some statically derived data for different grades (FE310, 360, 430, and 510 of Euronorm 25-1972 structural steel.)" He also reported a smoothed curve from unpublished work from Schleich of Arbed steel, of an unknown structural steel in the temperature range $(91 < T < 1135)$ °C.

Date   Date [70], from the British Steel Corporation, measured the elastic modulus of low-carbon steel (C=0.12, Mn=0.47, Si=0.16) at 400 hz using a resonance technique in the range $(20 < T < 600)$ °C. Although the data of Date [70] and Cooke [8] both originated from the British Steel Corporation, they diverge significantly for $T > 300$ °C.

Garofalo   Garofalo [71, 72] reported elastic modulus data measured in his laboratory obtained using simultaneous quasi-static bending and torsion of SAE 1015 steel at stresses less than 28 MPa in the range $(22 < T < 482)$ °C. Both papers also contain data for SAE 1015 steel that he describes as coming from unpublished investigations by the Steel and Tube division of the Timken Roller Bearing Co, during 1948–1950. He describes the measurements as "dynamic," but provides no other information.

Köster   Köster [73] measured the elastic modulus of pure metals including iron in the temperature range $(-150 < T < 1000)$ °C using a dynamic technique. His measurements are unique in that they extend into the austenite field of the iron phase diagram.

Ritter   As part of a study of stress-relaxation of welds, Ritter et al. [74] measured the temperature dependence of the stress-strain and elastic modulus of an Australian niobium-strengthened structural steel AS A151 in the temperature range $(20 < T < 572)$ °C. An interesting aspect of the study was that they tested both an as-normalized plate and a plate heat-treated to produce a larger grain size. The temperature dependence of the modulus of the two steels was significantly different, and demonstrates that anelastic effects can be significant and that grade and chemistry may not predict behavior. See Figure 33, for Reference 74. They attributed this difference to grain-boundary relaxation effects.

Roberts   Roberts and Nortcliffe [75] measured the elastic modulus of 18 steels, including a "mild steel" in the temperature range $(20 < T < 600)$ °C using a resonance technique at approximately 400 hz.

98

Uddin   Uddin and Culver [76] reported values for elastic modulus measured using unknown techniques on steels of unknown origin by Tall (1961) and by Stanzak (date and steel unknown) in the temperature range $(93 < T < 649)$ °C. The data of Tall come from an unavailable report. The data of Stanzak are mis-cited, and cannot be located.

Table 13: Data for elastic modulus of steels from literature sources.

Reference	Page	Author	Steel	T	E
				°C	GPa
[68]	93	ASME	[C]<0.3%	-198	216
[68]	93	ASME	[C]<0.3%	-129	212
[68]	93	ASME	[C]<0.3%	-73	209
[68]	93	ASME	[C]<0.3%	21	203
[68]	93	ASME	[C]<0.3%	93	199
[68]	93	ASME	[C]<0.3%	149	195
[68]	93	ASME	[C]<0.3%	204	192
[68]	93	ASME	[C]<0.3%	260	188
[68]	93	ASME	[C]<0.3%	316	183
[68]	93	ASME	[C]<0.3%	371	176
[68]	93	ASME	[C]<0.3%	427	167
[68]	93	ASME	[C]<0.3%	482	155
[68]	93	ASME	[C]<0.3%	538	141
[68]	93	ASME	[C]<0.3%	593	124
[68]	93	ASME	[C]>0.3%	-198	215
[68]	93	ASME	[C]>0.3%	-129	211
[68]	93	ASME	[C]>0.3%	-73	208
[68]	93	ASME	[C]>0.3%	21	201
[68]	93	ASME	[C]>0.3%	93	197
[68]	93	ASME	[C]>0.3%	149	194
[68]	93	ASME	[C]>0.3%	204	191
[68]	93	ASME	[C]>0.3%	260	187
[68]	93	ASME	[C]>0.3%	316	182
[68]	93	ASME	[C]>0.3%	371	174
[68]	93	ASME	[C]>0.3%	427	166
[68]	93	ASME	[C]>0.3%	482	154
[68]	93	ASME	[C]>0.3%	538	139
[68]	93	ASME	[C]>0.3%	593	123

continued on next page

Table 13: Data for elastic modulus of steels from literature sources.

Continued from the previous page

Reference	Page	Author	Steel	T	E
				°C	GPa
[68]	93	ASME	[C]>0.3%	649	106
[58]	96	USS	all structural	93	194
[58]	96	USS	all structural	204	188
[58]	96	USS	all structural	316	179
[58]	96	USS	all structural	371	175
[58]	96	USS	all structural	427	167
[58]	96	USS	all structural	482	158
[58]	96	USS	all structural	538	140
[69]	97	Clark	SAE1015	23	212
[69]	97	Clark	SAE1015	95	209
[69]	97	Clark	SAE1015	205	202
[69]	97	Clark	SAE1015	425	184
[69]	97	Clark	SAE1015	595	166
[69]	97	Clark	SAE1015	650	157
[8]	97	Cooke	Arbed	91	212
[8]	97	Cooke	Arbed	186	209
[8]	97	Cooke	Arbed	247	204
[8]	97	Cooke	Arbed	298	197
[8]	97	Cooke	Arbed	355	188
[8]	97	Cooke	Arbed	406	178
[8]	97	Cooke	Arbed	448	166
[8]	97	Cooke	Arbed	497	152
[8]	97	Cooke	Arbed	541	135
[8]	97	Cooke	Arbed	598	113
[8]	97	Cooke	Arbed	645	94
[8]	97	Cooke	Arbed	713	72
[8]	97	Cooke	Arbed	772	55
[8]	97	Cooke	Arbed	862	34
[8]	97	Cooke	Arbed	928	21
[8]	97	Cooke	Arbed	1023	9
[8]	97	Cooke	Arbed	1135	2
[8]	97	Cooke	Euronorm 25-1972	0	212
[8]	97	Cooke	Euronorm 25-1972	94	213

continued on next page

Table 13: Data for elastic modulus of steels from literature sources.

Continued from the previous page

Reference	Page	Author	Steel	T	E
				°C	GPa
[8]	97	Cooke	Euronorm 25-1972	135	203
[8]	97	Cooke	Euronorm 25-1972	230	192
[8]	97	Cooke	Euronorm 25-1972	287	184
[8]	97	Cooke	Euronorm 25-1972	358	169
[8]	97	Cooke	Euronorm 25-1972	415	156
[8]	97	Cooke	Euronorm 25-1972	473	140
[8]	97	Cooke	Euronorm 25-1972	530	122
[8]	97	Cooke	Euronorm 25-1972	584	104
[8]	97	Cooke	Euronorm 25-1972	621	89
[8]	97	Cooke	Euronorm 25-1972	667	68
[8]	97	Cooke	Euronorm 25-1972	728	48
[8]	97	Cooke	Euronorm 25-1972	778	34
[8]	97	Cooke	Euronorm 25-1972	793	31
[70]	97	Date	[C]=0.12 %	20	210
[70]	97	Date	[C]=0.12 %	100	204
[70]	97	Date	[C]=0.12 %	200	198
[70]	97	Date	[C]=0.12 %	300	190
[70]	97	Date	[C]=0.12 %	400	182
[70]	97	Date	[C]=0.12 %	500	172
[70]	97	Date	[C]=0.12 %	600	161
[71]	97	Garafolo	SAE1015	24	202
[71]	97	Garafolo	SAE1015	149	193
[71]	97	Garafolo	SAE1015	260	184
[71]	97	Garafolo	SAE1015	371	177
[71]	97	Garafolo	SAE1015	482	154
[72]	97	Garafolo	Timken Co.	22	212
[72]	97	Garafolo	Timken Co.	93	208
[72]	97	Garafolo	Timken Co.	149	205
[72]	97	Garafolo	Timken Co.	204	201
[72]	97	Garafolo	Timken Co.	260	196
[72]	97	Garafolo	Timken Co.	316	192
[72]	97	Garafolo	Timken Co.	371	188
[72]	97	Garafolo	Timken Co.	427	183

continued on next page

Table 13: Data for elastic modulus of steels from literature sources.

Continued from the previous page

Reference	Page	Author	Steel	T	E
				°C	GPa
[72]	97	Garafolo	Timken Co.	482	179
[72]	97	Garafolo	Timken Co.	538	174
[72]	97	Garafolo	Timken Co.	593	170
[72]	97	Garafolo	Timken Co.	649	161
[72]	97	Garafolo	Timken Co.	704	151
[72]	97	Garafolo	Timken Co.	760	141
[73]	97	Koester	Austenite	905	120
[73]	97	Koester	Austenite	912	120
[73]	97	Koester	Austenite	930	118
[73]	97	Koester	Austenite	935	117
[73]	97	Koester	Austenite	965	114
[73]	97	Koester	Austenite	985	112
[73]	97	Koester	Ferrite	20	211
[73]	97	Koester	Ferrite	105	206
[73]	97	Koester	Ferrite	245	196
[73]	97	Koester	Ferrite	390	186
[73]	97	Koester	Ferrite	498	176
[73]	97	Koester	Ferrite	600	167
[73]	97	Koester	Ferrite	680	157
[73]	97	Koester	Ferrite	715	147
[73]	97	Koester	Ferrite	720	145
[74]	97	Ritter	AS A151 ASTM GS 5	20	210
[74]	97	Ritter	AS A151 ASTM GS 5	98	205
[74]	97	Ritter	AS A151 ASTM GS 5	147	204
[74]	97	Ritter	AS A151 ASTM GS 5	254	195
[74]	97	Ritter	AS A151 ASTM GS 5	300	191
[74]	97	Ritter	AS A151 ASTM GS 5	348	182
[74]	97	Ritter	AS A151 ASTM GS 5	356	178
[74]	97	Ritter	AS A151 ASTM GS 5	424	175
[74]	97	Ritter	AS A151 ASTM GS 5	499	165
[74]	97	Ritter	AS A151 ASTM GS 5	548	162
[74]	97	Ritter	AS A151 ASTM GS 5	571	160
[74]	97	Ritter	AS A151 ASTM GS 5	572	157

continued on next page

Table 13: Data for elastic modulus of steels from literature sources.

Continued from the previous page

Reference	Page	Author	Steel	T	E
				°C	GPa
[74]	97	Ritter	AS A151 ASTM GS 8-9	20	210
[74]	97	Ritter	AS A151 ASTM GS 8-9	147	200
[74]	97	Ritter	AS A151 ASTM GS 8-9	149	202
[74]	97	Ritter	AS A151 ASTM GS 8-9	199	196
[74]	97	Ritter	AS A151 ASTM GS 8-9	298	173
[74]	97	Ritter	AS A151 ASTM GS 8-9	299	178
[74]	97	Ritter	AS A151 ASTM GS 8-9	299	175
[74]	97	Ritter	AS A151 ASTM GS 8-9	350	157
[74]	97	Ritter	AS A151 ASTM GS 8-9	350	152
[74]	97	Ritter	AS A151 ASTM GS 8-9	350	151
[74]	97	Ritter	AS A151 ASTM GS 8-9	350	147
[74]	97	Ritter	AS A151 ASTM GS 8-9	400	152
[74]	97	Ritter	AS A151 ASTM GS 8-9	400	147
[74]	97	Ritter	AS A151 ASTM GS 8-9	401	142
[74]	97	Ritter	AS A151 ASTM GS 8-9	448	148
[74]	97	Ritter	AS A151 ASTM GS 8-9	448	148
[74]	97	Ritter	AS A151 ASTM GS 8-9	449	144
[74]	97	Ritter	AS A151 ASTM GS 8-9	498	144
[74]	97	Ritter	AS A151 ASTM GS 8-9	499	143
[74]	97	Ritter	AS A151 ASTM GS 8-9	499	146
[74]	97	Ritter	AS A151 ASTM GS 8-9	500	137
[74]	97	Ritter	AS A151 ASTM GS 8-9	550	139
[74]	97	Ritter	AS A151 ASTM GS 8-9	550	143
[74]	97	Ritter	AS A151 ASTM GS 8-9	550	140
[75]	97	Roberts	mild steel	20	208
[75]	97	Roberts	mild steel	200	196
[75]	97	Roberts	mild steel	400	182
[75]	97	Roberts	mild steel	600	161
[76]	98	Uddin	not reported	93	200
[76]	98	Uddin	not reported	204	184
[76]	98	Uddin	not reported	260	170
[76]	98	Uddin	not reported	316	152
[76]	98	Uddin	not reported	371	132

continued on next page

Table 13: Data for elastic modulus of steels from literature sources.

Continued from the previous page

Reference	Page	Author	Steel	T	E
				°C	GPa
[76]	98	Uddin	not reported	427	114
[76]	98	Uddin	not reported	482	99
[76]	98	Uddin	not reported	538	86
[76]	98	Uddin	not reported	649	66
[76]	98	Uddin	structural	93	200
[76]	98	Uddin	structural	204	193
[76]	98	Uddin	structural	260	186
[76]	98	Uddin	structural	316	182
[76]	98	Uddin	structural	371	175
[76]	98	Uddin	structural	427	167
[76]	98	Uddin	structural	482	159
[76]	98	Uddin	structural	538	146
[76]	98	Uddin	structural	649	121

E.5 Strain rate sensitivity of literature steels

Figure 5 combines data from several of the literature sources. [6, 25, 29, 31, 33, 40, 62] Where possible, the strain rate sensitivity, m, for each of the steels in these sources was calculated from the reported strain rates and 0.2 % offset yield strengths, S_y(0.002 offset), summarized in Table 10. At each temperature, a linear regression of that yield strength, $\log_e(S_y)$, on the natural logarithm of the strain rate, $\log_e(\dot{\epsilon})$ produced an estimate of the strain rate sensitivity, m. If the data contained more than two points at that temperature, the regression also yields an estimate of the uncertainty of the strain rate sensitivity, $u(m)$. Table 8, at the end of this section, compiles the computed strain rate sensitivities for each steel.

Figure 36 shows the data used to calculate the strain rate sensitivity, m, for the steels listed in Table 15, grouped by temperature range.

E.5.1 A model for $m(T)$ for structural steel

Figure 5 plots a fit of Eq. (9) to the entire data set including the steels of this report and the literature steels that the previous section describes. Table 3 summarizes the parameters. Like other non-linear least-squares fits to data in this report, the fit was constrained:

m_0 not fit,
$1 \leq m_1 \leq 10$,

104

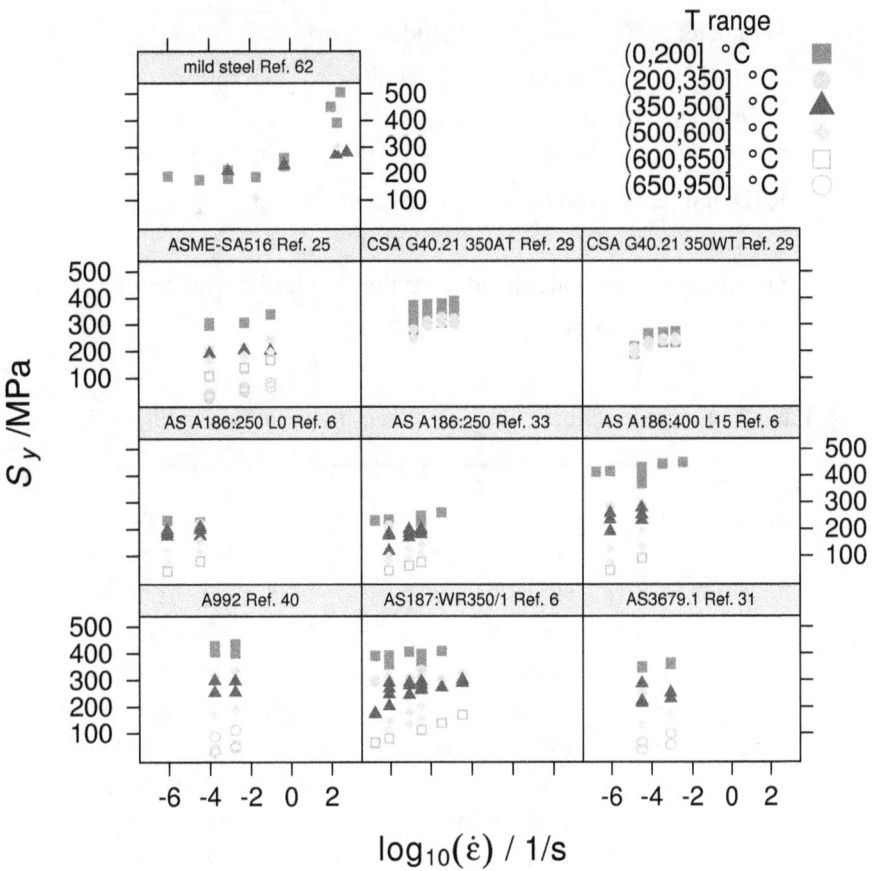

Figure 36: 0.2 % offset yield strength, $S_y(0.2\,\%$ offset), as a function of strain rate, $\dot{\epsilon}$ for literature data. Data are grouped by temperature range.

$(100 \leq m_2 \leq 3000)\,°\text{C}$,
$0.01 \leq m_3 \leq 0.25$,
$T \geq 390\,°\text{C}$,
data from Ref. 62 omitted.

The strain rate sensitivity at room temperature, m_0, was not fit. Instead it was set to the mean value of the room-temperature data. None of the values of the parameters in Table 14 are near the constraints.

Table 14: Output of the regression for $m(T)$.

Parameter	Estimate	Standard Error	t
m_1	7.30844	1.54486	4.73082
m_2	613.10509	21.78958	28.13754
m_3	0.12602	0.01321	9.54318
RSD: 0.03259 on 68 degrees of freedom			
m_0	0.01079	0.00261	

uncertainty of m_0 calculated from the standard deviation of 23 measurements of m_0.

Table 15: Literature values of strain rate sensitivity m used in Figure 5.

Reference	Section	Steel	T °C	m	$u(m)$
[25]	E.1.16	ASME-SA516	25	0.01512	0.00548
[25]	E.1.16	ASME-SA516	315	0.02204	0.00305
[25]	E.1.16	ASME-SA516	427	0.00989	0.00319
[25]	E.1.16	ASME-SA516	537	0.02574	0.00281
[25]	E.1.16	ASME-SA516	649	0.06432	0.0017
[25]	E.1.16	ASME-SA516	760	0.10941	0.0072
[25]	E.1.16	ASME-SA516	815	NA	NA
[25]	E.1.16	ASME-SA516	871	0.0878	0.00542
[25]	E.1.16	ASME-SA516	927	NA	NA
[62]	E.1.16	mild steel	20	0.05047	0.01083
[62]	E.1.16	mild steel	200	0.05142	0.01211
[62]	E.1.16	mild steel	400	0.02178	0.00204
[62]	E.1.16	mild steel	600	0.11158	0.00256
[31]	E.1.16	AS3679.1	20	0.01195	0.00294
[31]	E.1.16	AS3679.1	300	NA	NA
[31]	E.1.16	AS3679.1	400	0.00089	0.07214
[31]	E.1.16	AS3679.1	500	0.01898	0.00877
[31]	E.1.16	AS3679.1	600	0.0869	0.0
[31]	E.1.16	AS3679.1	700	0.12675	0.00416
[31]	E.1.16	AS3679.1	800	0.11937	NA

NA: not available

continued on next page

Table 15: Literature values of strain rate sensitivity m used in Figure 5.

Continued from the previous page

Reference	Steel	Steel	T °C	m	$u(m)$
[33]	E.1.16	AS A186:250	27	0.01645	0.00133
[33]	E.1.16	AS A186:250	100	NA	NA
[33]	E.1.16	AS A186:250	200	NA	NA
[33]	E.1.16	AS A186:250	300	-0.01565	NA
[33]	E.1.16	AS A186:250	350	0.0033	0.01495
[33]	E.1.16	AS A186:250	400	0.02776	0.00679
[33]	E.1.16	AS A186:250	450	0.02429	0.004
[33]	E.1.16	AS A186:250	500	0.11788	0.0291
[33]	E.1.16	AS A186:250	550	0.10429	0.00573
[33]	E.1.16	AS A186:250	600	0.07879	0.01795
[33]	E.1.16	AS A186:250	650	0.14569	0.00035
[6]	E.1.16	AS187:WR350/1	27	0.00562	0.00177
[6]	E.1.16	AS187:WR350/1	100	0.00288	NA
[6]	E.1.16	AS187:WR350/1	200	0.00743	0.0013
[6]	E.1.16	AS187:WR350/1	300	0.02077	NA
[6]	E.1.16	AS187:WR350/1	350	0.00335	0.00267
[6]	E.1.16	AS187:WR350/1	400	0.00415	0.00326
[6]	E.1.16	AS187:WR350/1	450	0.01261	0.00215
[6]	E.1.16	AS187:WR350/1	475	0.02304	NA
[6]	E.1.16	AS187:WR350/1	500	0.05179	0.00583
[6]	E.1.16	AS187:WR350/1	550	0.08645	0.00193
[6]	E.1.16	AS187:WR350/1	600	0.08872	0.00039
[6]	E.1.16	AS187:WR350/1	650	0.09698	0.00316
[6]	E.1.16	AS A186:250 L0	27	-0.0071	NA
[6]	E.1.16	AS A186:250 L0	350	0.03814	NA
[6]	E.1.16	AS A186:250 L0	400	0.01695	NA
[6]	E.1.16	AS A186:250 L0	450	0.01741	NA
[6]	E.1.16	AS A186:250 L0	500	0.0031	NA
[6]	E.1.16	AS A186:250 L0	550	0.07287	NA
[6]	E.1.16	AS A186:250 L0	600	0.12696	NA
[6]	E.1.16	AS A186:250 L0	650	0.16832	NA
[6]	E.1.16	AS A186:400 L15	27	0.00905	0.00075

NA: not available

continued on next page

Table 15: Literature values of strain rate sensitivity m used in Figure 5.

Continued from the previous page

Reference	Steel	Steel	T	m	$u(m)$
			°C		
[6]	E.1.16	AS A186:400 L15	100	NA	NA
[6]	E.1.16	AS A186:400 L15	200	NA	NA
[6]	E.1.16	AS A186:400 L15	300	0.00563	NA
[6]	E.1.16	AS A186:400 L15	350	0.00849	NA
[6]	E.1.16	AS A186:400 L15	400	0.02009	NA
[6]	E.1.16	AS A186:400 L15	450	0.01993	NA
[6]	E.1.16	AS A186:400 L15	500	0.05389	NA
[6]	E.1.16	AS A186:400 L15	550	0.1142	NA
[6]	E.1.16	AS A186:400 L15	600	0.15622	NA
[6]	E.1.16	AS A186:400 L15	650	0.19389	NA
[40]	E.1.16	A992	20	0.00703	NA
[40]	E.1.16	A992	100	-0.0032	NA
[40]	E.1.16	A992	200	-0.00538	NA
[40]	E.1.16	A992	300	0.02283	NA
[40]	E.1.16	A992	400	-0.00146	NA
[40]	E.1.16	A992	500	0.00171	NA
[40]	E.1.16	A992	600	0.04804	NA
[40]	E.1.16	A992	700	0.10474	NA
[40]	E.1.16	A992	800	0.16419	NA
[40]	E.1.16	A992	900	0.22185	NA
[29]	E.1.16	CSA G40.21 350AT	20	0.00806	0.00194
[29]	E.1.16	CSA G40.21 350AT	100	0.00927	0.00144
[29]	E.1.16	CSA G40.21 350AT	150	0.01705	0.00182
[29]	E.1.16	CSA G40.21 350AT	200	0.02347	0.00883
[29]	E.1.16	CSA G40.21 350AT	250	0.02887	0.01191
[29]	E.1.16	CSA G40.21 350AT	300	0.04061	0.01845
[29]	E.1.16	CSA G40.21 350WT	20	0.04512	0.02008
[29]	E.1.16	CSA G40.21 350WT	100	0.04133	0.01083
[29]	E.1.16	CSA G40.21 350WT	150	0.04919	0.02241
[29]	E.1.16	CSA G40.21 350WT	200	0.02672	0.01558
[29]	E.1.16	CSA G40.21 350WT	250	0.03618	0.0104
[29]	E.1.16	CSA G40.21 350WT	300	0.02649	0.0117

NA: not available

continued on next page

Table 15: Literature values of strain rate sensitivity m used in Figure 5.

Continued from the previous page					
Reference	Steel	Steel	T °C	m	$u(m)$
[29]	E.1.16	CSA G40.21 350WT	350	0.04902	0.01298

NA: not available

F   Eurocode stress-strain model

No publicly available document explains the basis for the choices of the form of the Eurocode 3 [1] stress-strain model or the values of the parameters in it. This section summarizes the Eurocode 3 stress-strain model and compares its predictions to the data of this report.

F.1   Shape of the stress-strain curve

The shape of the *engineering* stress-strain, $S - e$, curve in the Eurocode 3 formulation comprises four regions:

1. a linear elastic region,

2. a plastic region described by an ellipse that connects the end of the linear-elastic region to $e = 0.02$,

3. a plastic region of constant maximum stress with increasing strain, possibly with a short region of strain hardening for $T < 400\,°\text{C}$, and

4. a plastic region where the stress decreases from the maximum to zero.

Figure 37 shows the shape of the engineering stress-strain curve in the Eurocode 3 model in regions I, II, and III. The temperature dependence of the engineering stress-strain behavior in the first three regions is captured in four parameters:

1. a temperature-dependent Young's modulus, $\bar{E}_{a,\theta}$,

2. a temperature-dependent proportional limit, $f_{\text{ap},\theta}$,

3. a temperature-dependent maximum stress, $f_{\text{amax},\theta}$, and

4. a temperature-dependent tensile strength with strain hardening, $f_{\text{au},\theta}$

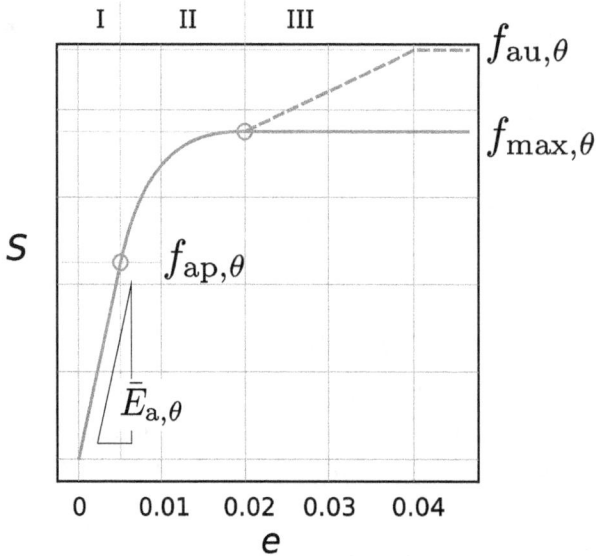

Figure 37: The shape of the Eurocode 3 stress-strain curve. Dashed line shows the shape of the curve for temperatures, $T \leq 400 \, ^\circ C$.

Three separate equations describe the change in engineering stress, S, in the first three regions of engineering strain, e:

$$S = \begin{cases} \bar{E}_{a,\theta} \, e & \text{for } 0 \leq e \leq \dfrac{f_{\text{ap},\theta}}{\bar{E}_{a,\theta}}, \text{ (Reg. 1)} \\[2ex] \dfrac{b}{a} \sqrt{a^2 + (0.02 - e)^2} + f_{\text{ap},\theta} - c & \text{for } \dfrac{f_{\text{ap},\theta}}{\bar{E}_{a,\theta}} \leq e \leq 0.02, \text{ (Reg. 2)} \\[2ex] f_{\text{amax},\theta} & \text{for } e \geq 0.02 \text{ and } T > 400 \, ^\circ C, \text{ (Reg. 3)} \end{cases}$$

$$(20)$$

For temperatures below $400 \, ^\circ C$, Region III is modified to express strain hardening. For these conditions

$$S = \begin{cases} \left(\dfrac{f_{\text{au},\theta} - f_{\text{amax},\theta}}{0.02} \right) e - f_{\text{au},\theta} + 2 f_{\text{amax},\theta} & \text{for } 0.02 < e < 0.04 \\[2ex] f_{\text{au},\theta} & \text{for } e \geq 0.04 \end{cases}$$

$$(21)$$

The parameters, a, b, and c that describe the ellipse that connects the elastic region (I) to the constant plastic region (III) are

$$a = \sqrt{(0.02 - \frac{f_{ap,\theta}}{\bar{\bar{E}}_{a,\theta}})(0.02 - \frac{f_{ap,\theta}}{\bar{\bar{E}}_{a,\theta}} + \frac{c}{\bar{\bar{E}}_{a,\theta}})} \tag{22}$$

$$b = \sqrt{\bar{E}_{a,\theta}(0.02 - \frac{f_{ap,\theta}}{\bar{\bar{E}}_{a,\theta}})c + c^2} \tag{23}$$

$$c = \frac{(f_{amax,\theta} - f_{ap,\theta})^2}{\bar{E}_{a,\theta}(0.02 - \frac{f_{ap,\theta}}{\bar{\bar{E}}_{a,\theta}}) - 2(f_{amax,\theta} - f_{ap,\theta})} \tag{24}$$

The temperature dependence of the three terms that describe the change in shape of the stress-strain curve, Figure 37, are expressed as discrete points connected by linear segments rather than by smooth functions. In addition, they are expressed as values normalized to either the room-temperature elastic modulus, in the case of $\bar{E}_{a,\theta}$, or the room-temperature yield strength, $S_y(0.2\,\%$ offset) in the case of $f_{ap,\theta}$, $f_{amax,\theta}$, and $f_{au,\theta}$:

$$k_{E,\theta} = \frac{\bar{E}_{a,\theta}}{\bar{\bar{E}}_{T=20\,°C}} \qquad \text{Modulus} \tag{25}$$

$$k_{p,\theta} = \frac{f_{ap,\theta}}{S_y} \qquad \text{Proportional limit} \tag{26}$$

$$k_{max,\theta} = \frac{f_{amax,\theta}}{S_y} \qquad \text{Maximum stress} \tag{27}$$

$$k_{u,\theta} = \frac{f_{au,\theta}}{S_y} \qquad \text{Tensile strength} \tag{28}$$

where $S_y = S_y(0.2\,\%$ offset), the room-temperature 0.2 % offset yield strength. It is important to recognize that neither the proportional limit, $f_{ap,\theta}$, nor the maximum stress, $f_{amax,\theta}$, in the Eurocode 3 formulation corresponds to points usually calculated in room- or elevated-temperature stress-strain curves. Figure 38 and Table 16 show the temperature dependence of the four parameters. Symbols in Figure 38 indicate the fixed points.

Figure 39 shows the evolution of the engineering stress-strain curve with temperature. Note that the individual curves for $(20 \leq T \leq 300)$ °C lie on top of each other for $e > 0.02$. Only the shape of the curve for $e < 0.02$ changes for those temperatures, as the knee at $T = 20$ °C transforms into a shallower ellipse.

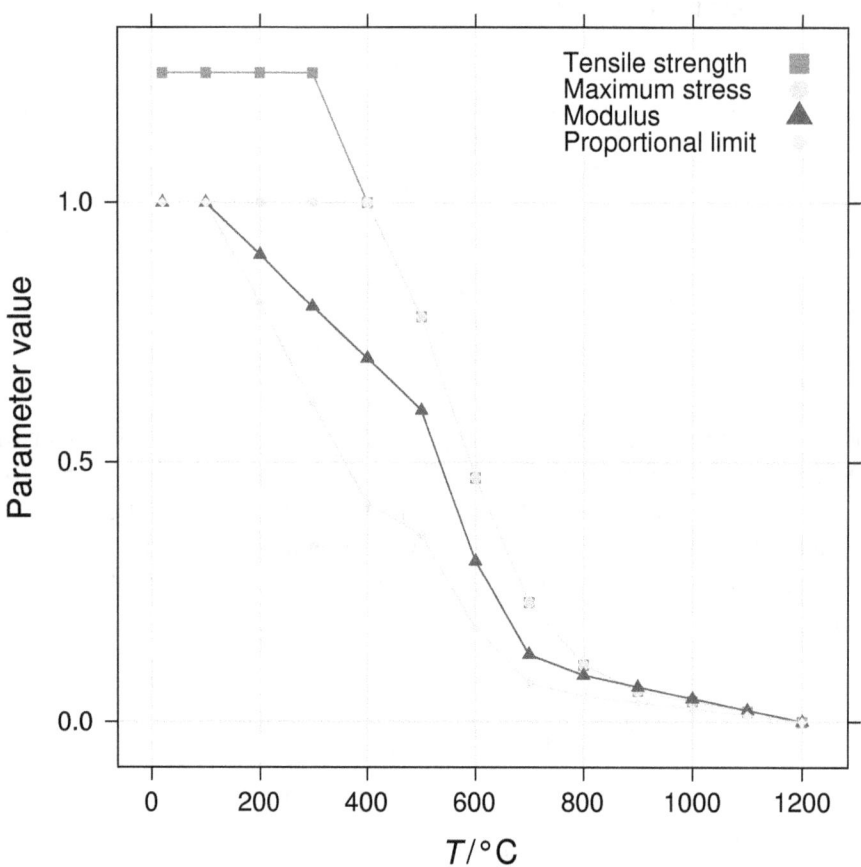

Figure 38: Temperature dependence of the four strength parameters in the Eurocode 3 stress-strain equation: tensile strength $f_{\mathrm{au},\theta}$, maximum stress $f_{\mathrm{amax},\theta}$, elastic modulus $\bar{E}_{a,\theta}$, and proportional limit $f_{\mathrm{ap},\theta}$. Equations (25), (26), (27), and (28) define the parameters.

F.2   Computing the yield strength

The Eurocode 3 stress-strain model does not contain an explicit expression for the traditional $e = 0.002$ offset yield strength, $S_y(0.002$ offset$)$, that typically appears in literature reports. It is possible, however to compute it for by solving numerically for the intersection of the offset elastic modulus line and the stress-strain curve.

112

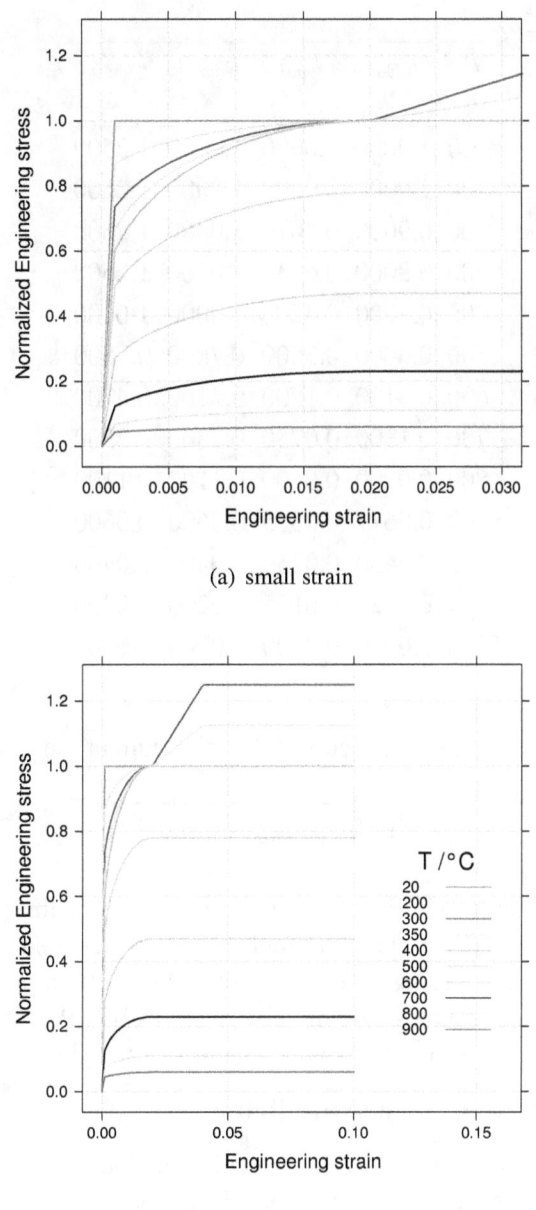

(a) small strain

(b) large strain

Figure 39: Temperature dependence of the Eurocode 3 engineering stress-strain curve. (a) small strain, and (b) large strain

Table 16: Eurocode 3 non-dimensional reduction factors used in Eq. (25), Eq. (26), Eq. (27), and Eq. (28).

T	$k_{E,\theta}$	$k_{p,\theta}$	$k_{\max,\theta}$	$k_{u,\theta}$
°C				
20	1.0000	1.0000	1.0000	1.2500
100	1.0000	1.0000	1.0000	1.2500
200	0.9000	0.8070	1.0000	1.2500
300	0.8000	0.6130	1.0000	1.2500
400	0.7000	0.4200	1.0000	1.0000
500	0.6000	0.3600	0.7800	0.7800
600	0.3100	0.1800	0.4700	0.4700
700	0.1300	0.0750	0.2300	0.2300
800	0.0900	0.0500	0.1100	0.1100
900	0.0675	0.0375	0.0600	0.0600
1000	0.0450	0.0250	0.0400	0.0400
1100	0.0225	0.0125	0.0200	0.0200
1200	0.0000	0.0000	0.0000	0.0000

The strain at the intersection, e', occurs at the minimum of the function

$$f = \sqrt{(S(e) - \bar{E}_{a,\theta}(e - 0.002))^2} \tag{29}$$

The stress, $S(e)$, is defined in Eq. (20) and Eq. (21), and the linear elastic portion, $\bar{E}_{a,\theta}$, is defined in Eq. (25) The yield strength is can be calculated from $S_y(e')$. This line is plotted in Figure 2, using $S_y(0.2 \% \text{ offset})$=250 MPa (approximately 36 ksi) and room-temperature Young's modulus $\bar{E}_{a,20 \, °C} = 210$ GPa. The absolute value of the retained strength, R, is a very weak function of the yield strength chosen.

F.3 Comparison to Eurocode prediction

Figure 40, which complements the relative deviation plot, Fig. 8, plots the absolute deviations of the stress-strain curves between the Eurocode prediction, Eq. (20) and the data of this study. Note that the Eurocode prediction for the five lowest-strength steels is consistently lower than the actual data. Figures 41 through 49 compare the prediction of the NIST global model, Eq. (10) and Table 3, to the prediction of the Eurocode 3 stress-strain model, Eq. (20) and (21). In the following figures, the prediction of the Eurocode 3 engineering stress-strain, $S - e$, model has been

114

transformed to true stress-strain $\sigma - \epsilon$ behavior by

$$\sigma = S(1+e) \qquad (30)$$

$$\epsilon = \log_e(1+e) \qquad (31)$$

For the NIST models of the individual steels, the only input is the measured room-temperature yield strength.

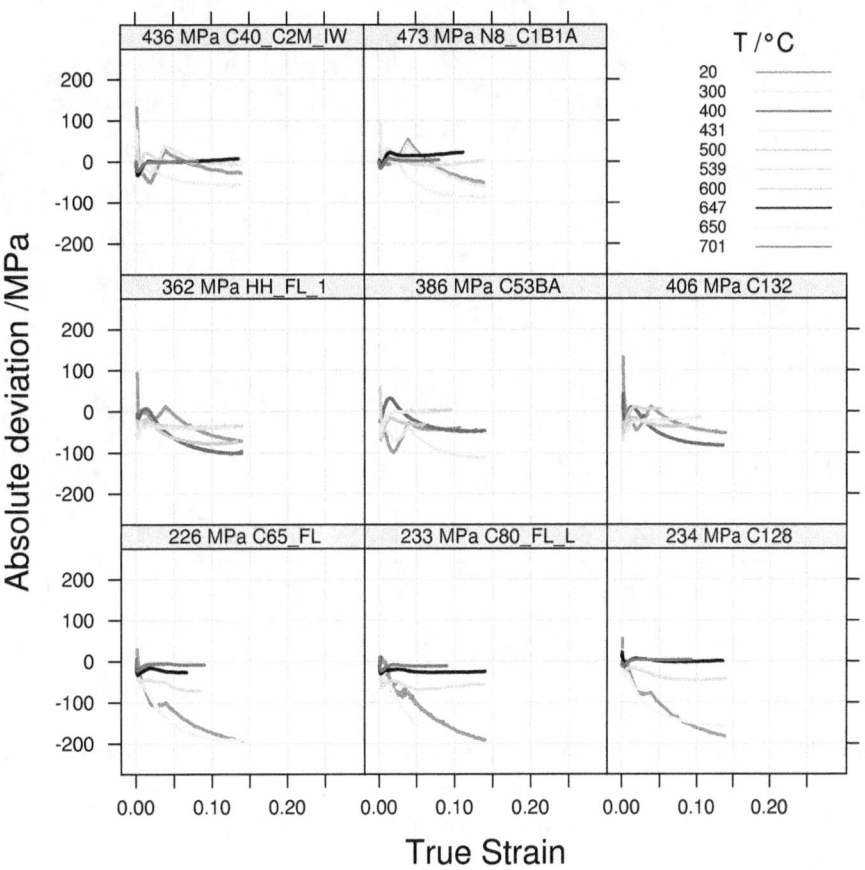

Figure 40: Absolute deviation, D_a between true stress predicted by the Eurocode 3 stress-strain model, Eq. (20), and the observed behavior.

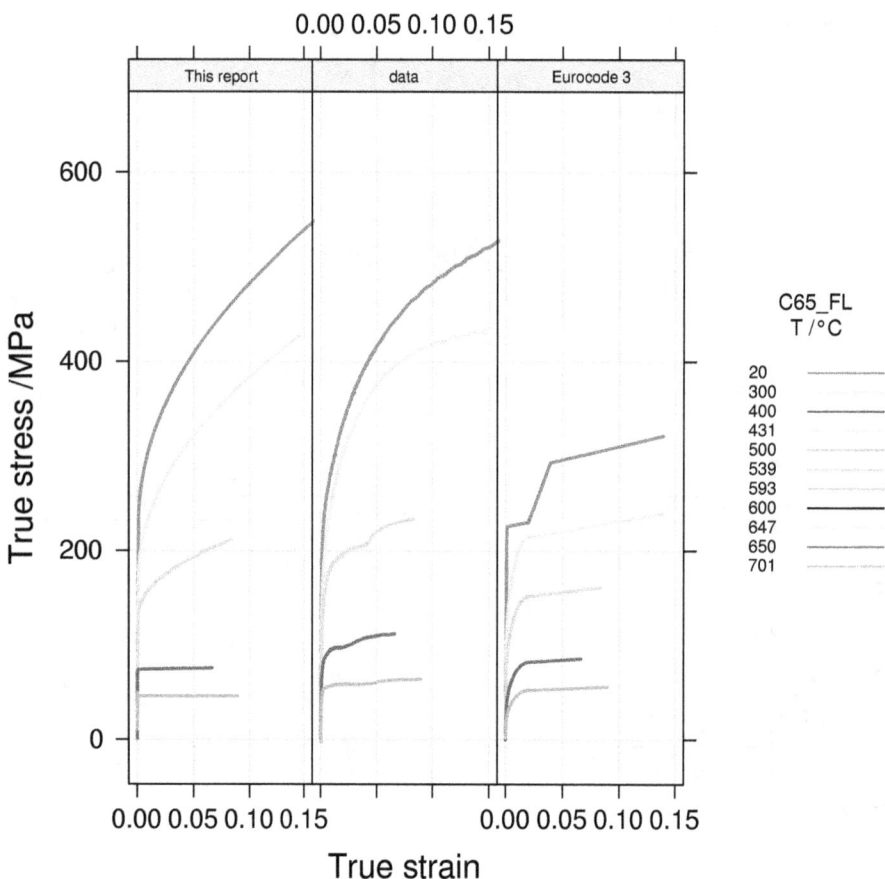

Figure 41: Comparison of predicted true stress, Eq. (10), to prediction of Eurocode 3 model for specimen C65_FL. $S_y(0.2\,\%\text{ offset}) = 226$ MPa. Left plot: this report; center plot: original data; right plot: Eurocode 3 prediction.

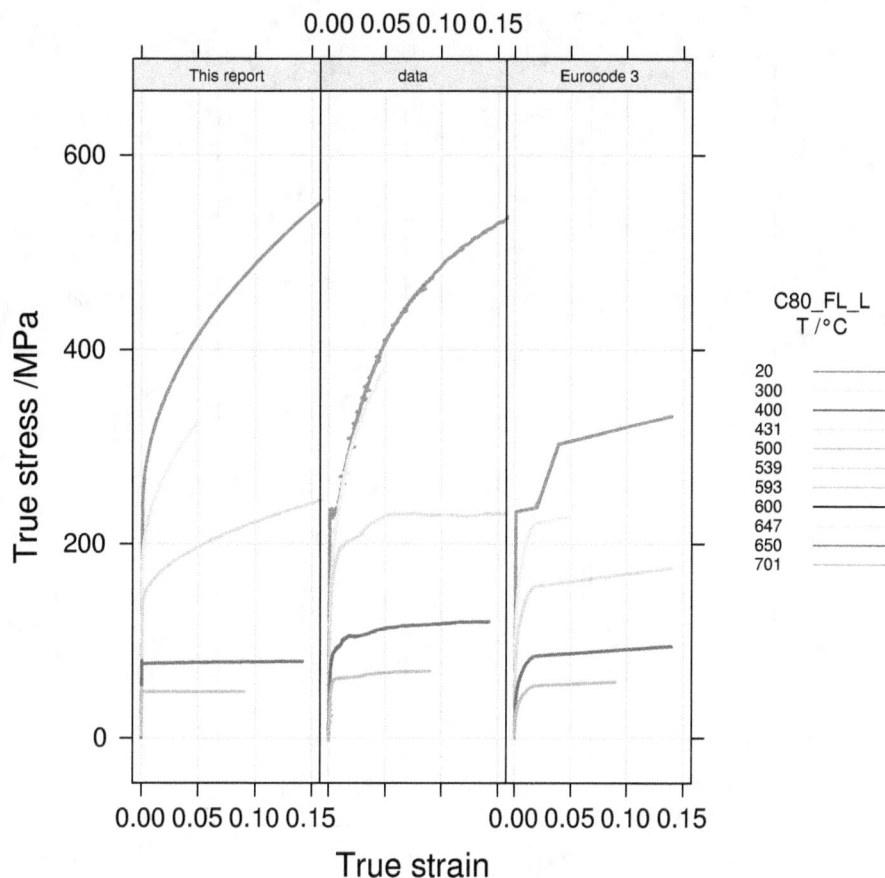

Figure 42: Comparison of predicted true stress, Eq. (10), to prediction of Eurocode 3 model for specimen C80_FL_L. $S_y(0.2\,\%\text{ offset}) = 233$ MPa. Left plot: this report; center plot: original data; right plot: Eurocode 3 prediction.

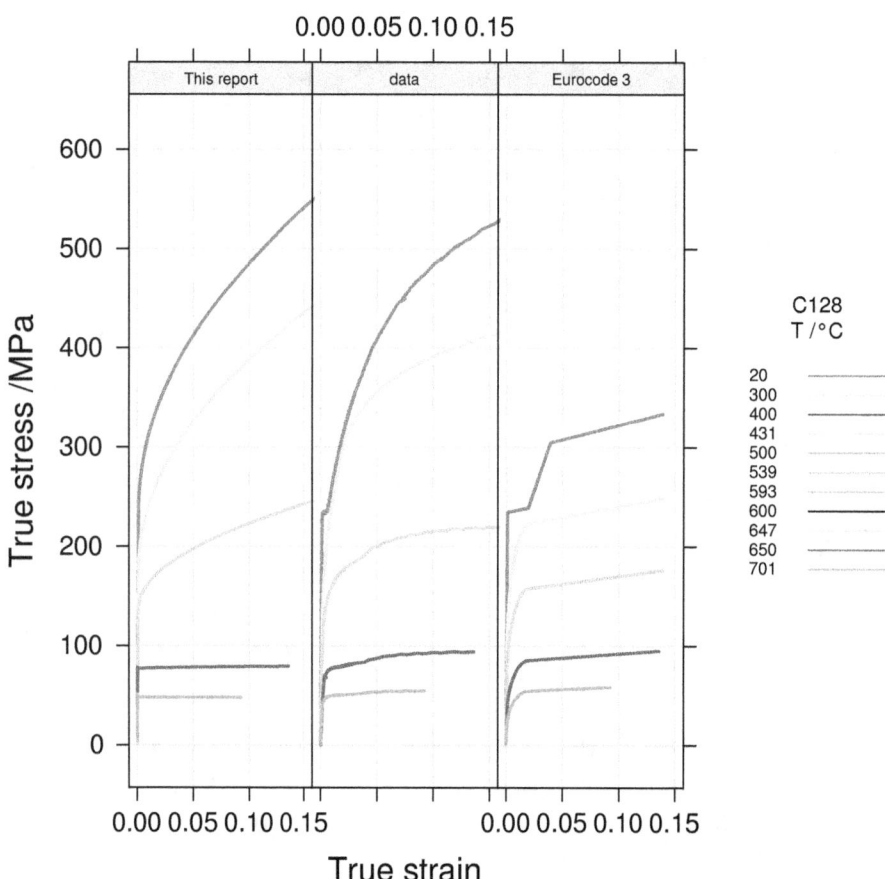

Figure 43: Comparison of predicted true stress, Eq. (10), to prediction of Eurocode 3 model for specimen C128. $S_y(0.2\,\%\,\text{offset}) = 234$ MPa. Left plot: this report; center plot: original data; right plot: Eurocode 3 prediction.

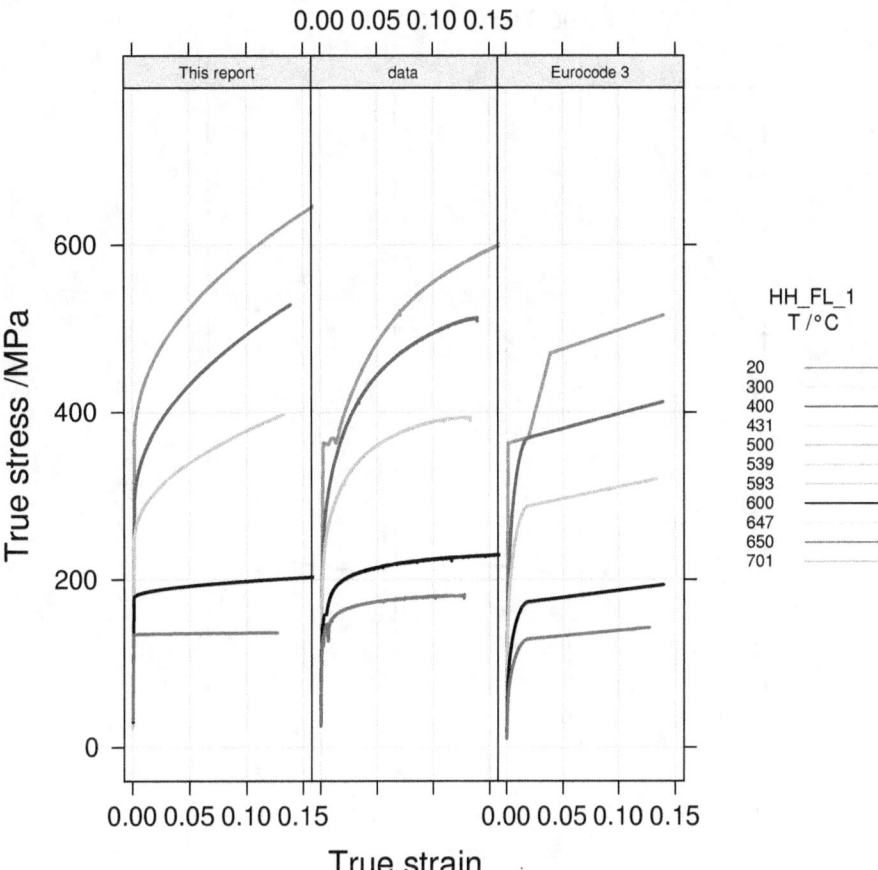

Figure 44: Comparison of predicted true stress, Eq. (10), to prediction of Eurocode 3 model for specimen HH_FL_1. $S_y(0.2\% \text{ offset}) = 362.1$ MPa. Left plot: this report; center plot: original data; right plot: Eurocode 3 prediction.

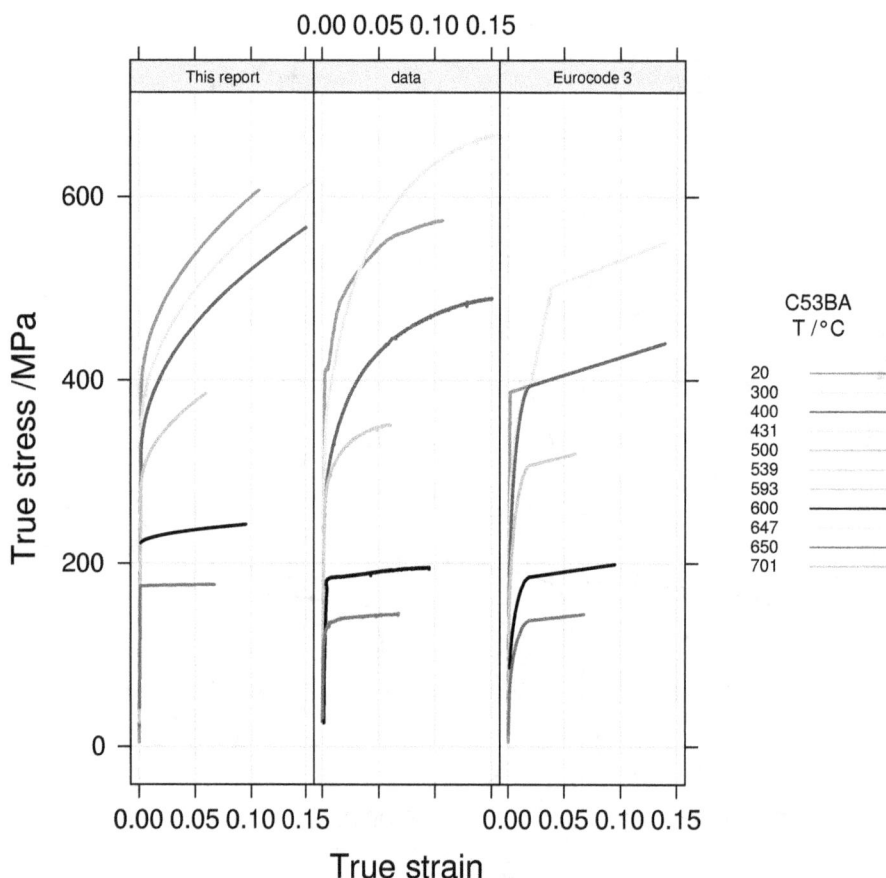

Figure 45: Comparison of predicted true stress, Eq. (10), to prediction of Eurocode 3 model for specimen C53BA. $S_y(0.2\,\%\,\mathrm{offset}) = 386$ MPa. Left plot: this report; center plot: original data; right plot: Eurocode 3 prediction.

120

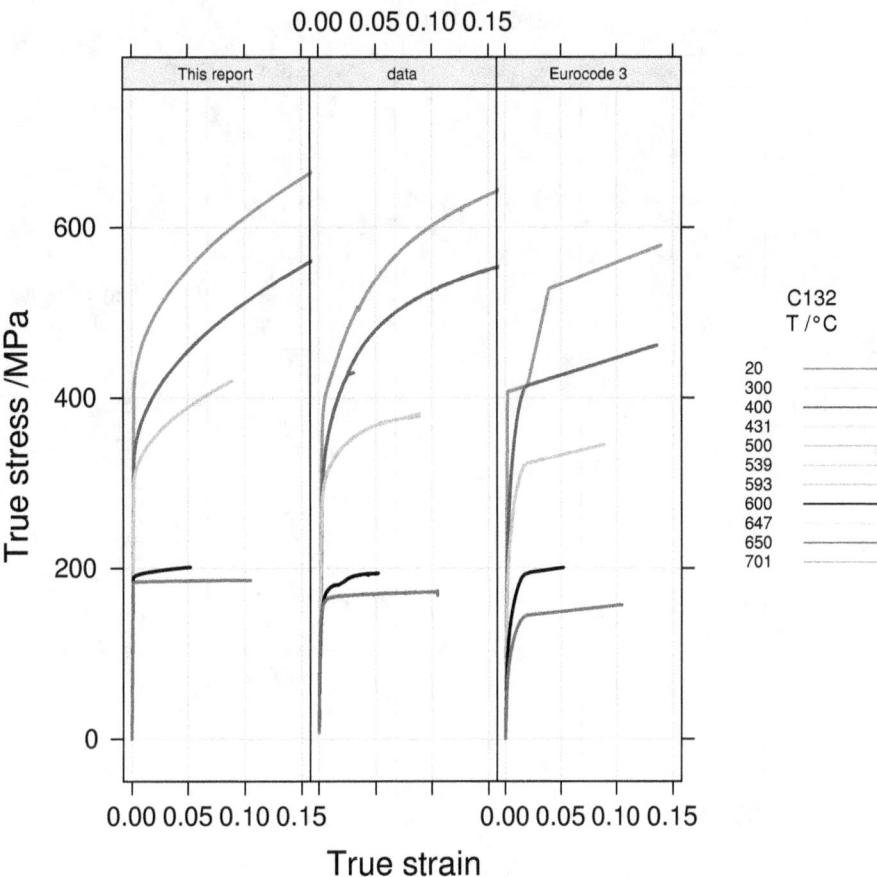

Figure 46: Comparison of predicted true stress, Eq. (10), to prediction of Eurocode 3 model for specimen C132. $S_y(0.2\% \text{ offset}) = 406$ MPa. Left plot: this report; center plot: original data; right plot: Eurocode 3 prediction.

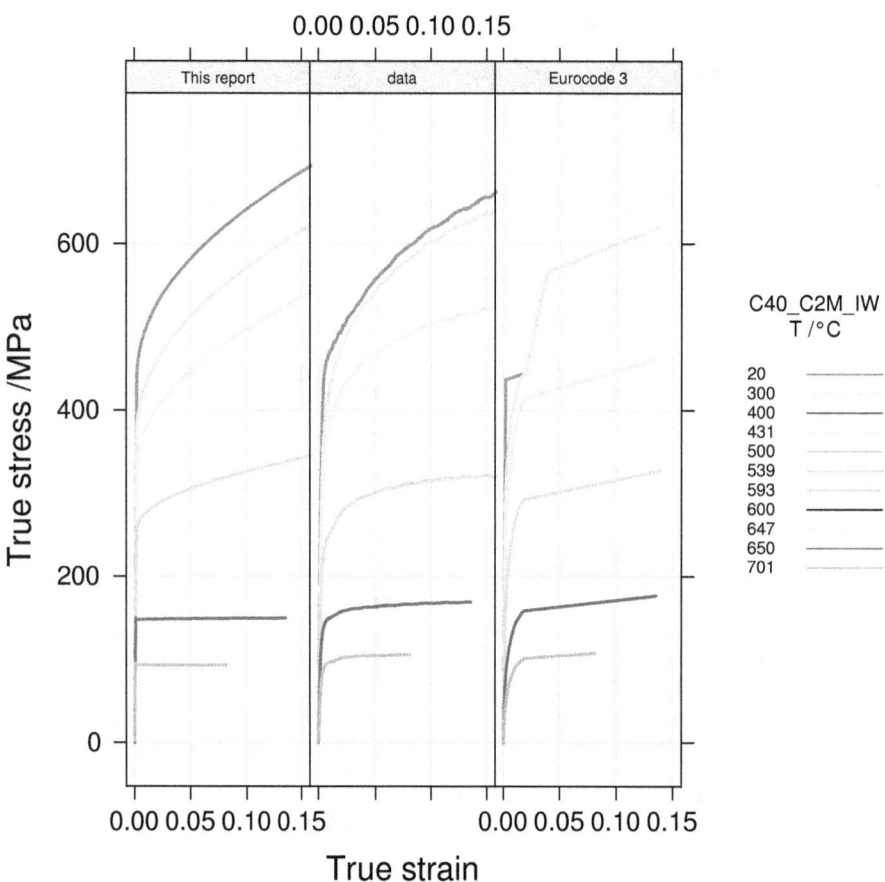

Figure 47: Comparison of predicted true stress, Eq. (10), to prediction of Eurocode 3 model for specimen C40_C2M_IW. $S_y(0.2\%\ \text{offset}) = 436$ MPa. Left plot: this report; center plot: original data; right plot: Eurocode 3 prediction.

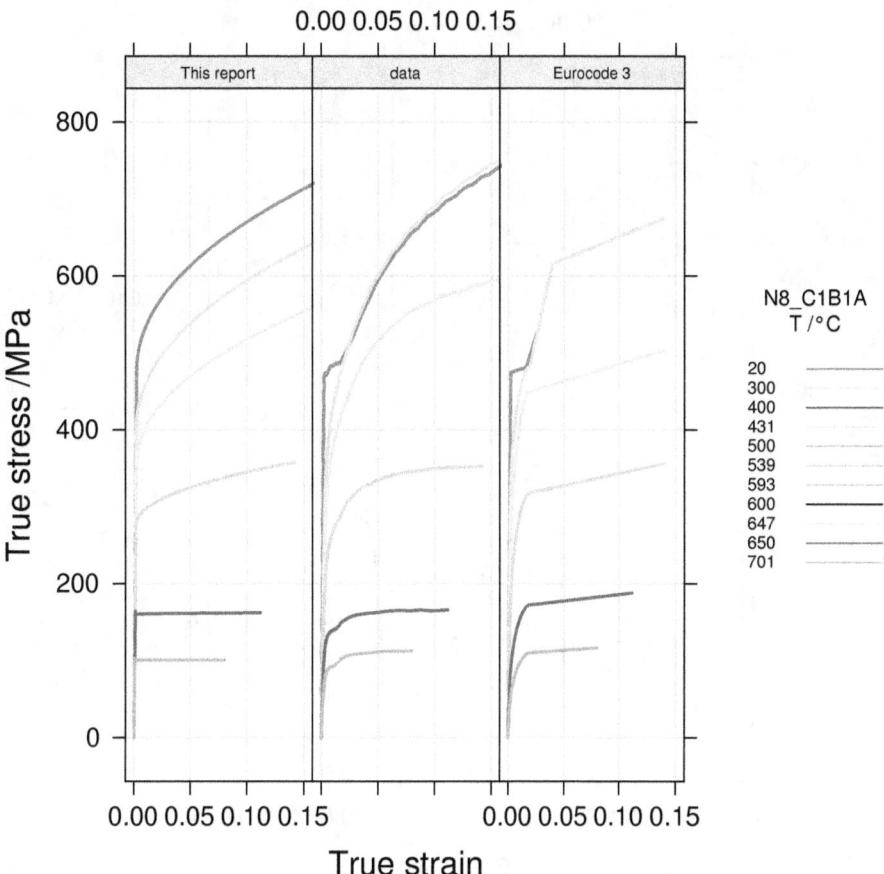

Figure 48: Comparison of predicted true stress, Eq. (10), to prediction of Eurocode 3 model for specimen N8_C1B1A. $S_y(0.2\,\%\text{ offset}) = 473.4$ MPa. Left plot: this report; center plot: original data; right plot: Eurocode 3 prediction.

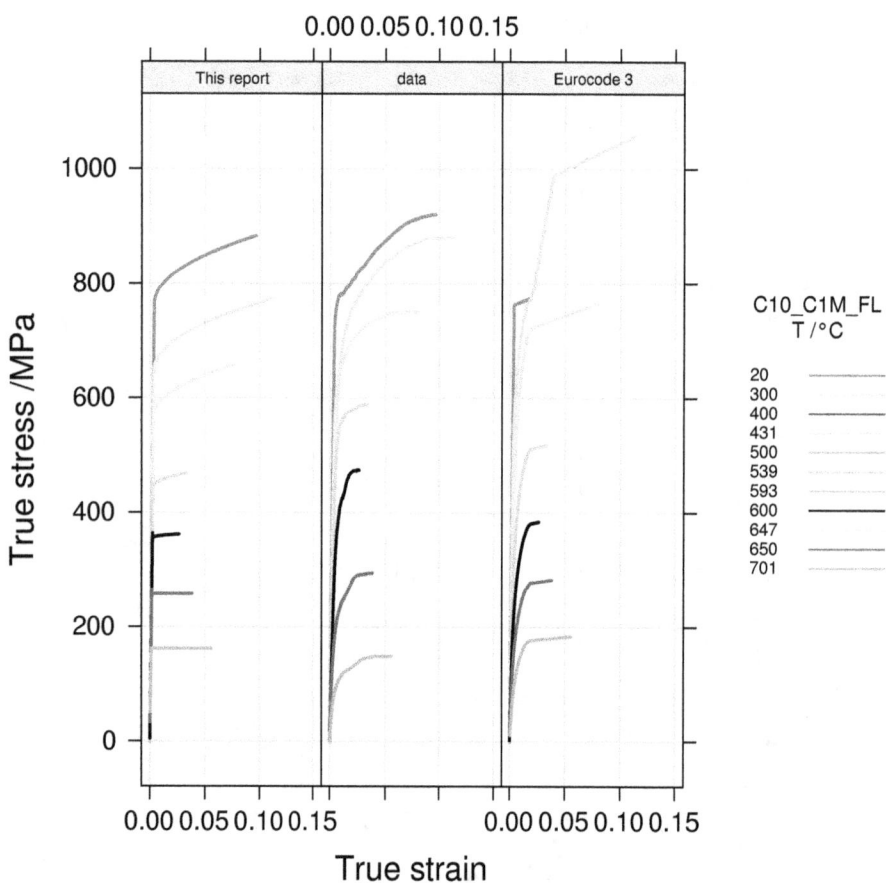

Figure 49: Comparison of predicted true stress, Eq. (10), to prediction of Eurocode 3 model for specimen C10_C1M_FL. $S_y(0.2 \% \text{ offset}) = 760$ MPa. Left plot: this report; center plot: original data; right plot: Eurocode 3 prediction.

G ECCS stress-strain model

The European Convention for Constructional Steelwork (ECCS) Technical Committee 3 published a recommended stress-strain model [9] for structural steel in 1983. Aside from a citations by some of the report's authors [39], the stress-strain model has not been frequently used. The model itself was "based on data obtained in the Netherlands and elsewhere." Unlike other models of the era, it includes the effects of time-dependent deformation through an effective yield strength that degrades with temperature faster than the measured yield strength. The report, however, does not explain the technical basis for the calculation of effective yield strength or support the model with any data. The reduction, R, in effective yield strength is

$$R = 1 + \frac{T}{t_{e1} \log_e \dfrac{T}{t_{e2}}} \quad 0\,°\text{C} \leq T \leq 600\,°\text{C} \tag{32}$$

$$R = \frac{t_{e3}\left(1 - \dfrac{T}{t_{e4}}\right)}{T - t_{e5}} \quad 600\,°\text{C} \leq T \leq 1000\,°\text{C}$$

where

$t_{e1} = 767\,°\text{C},$
$t_{e2} = 1750\,°\text{C},$
$t_{e3} = 108,$
$t_{e4} = 1000\,°\text{C, and}$
$t_{e5} = 440\,°\text{C}.$

Figure 50 compares the retained strength from Eq. (32) to the other retained strength models, including the Eurocode 3 [1], section F.2 and the four-parameter function of this report, Eq. (1). The ECCS model predicts the lowest retained strength, which is consistent with its assertion that it incorporates time-dependent deformation. The retained strength line lies well below all the other models for $T = 400\,°\text{C}$, where creep is not significant, however.

The ECCS report does not describe the stress-strain behavior analytically, but instead displays stress-strain curves for discrete temperatures graphically as Fig. r-1. Figure 51 compares the the Eurocode 3 [1] and the ECCS stress-strain models digitized from the report in the range $0 < e < 0.04$, for a steel with specified yield strength, $F_y = 235$ MPa. The stress-strain curves strain harden briefly after yield to the effective yield strain, which typically occurs $e < 0.005$, and then become perfectly plastic. The individual points on the stress-strain curves are not calculated

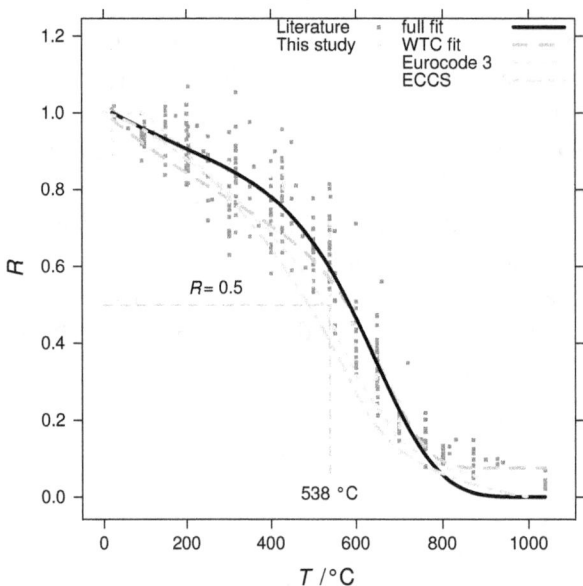

(a) full temperature range

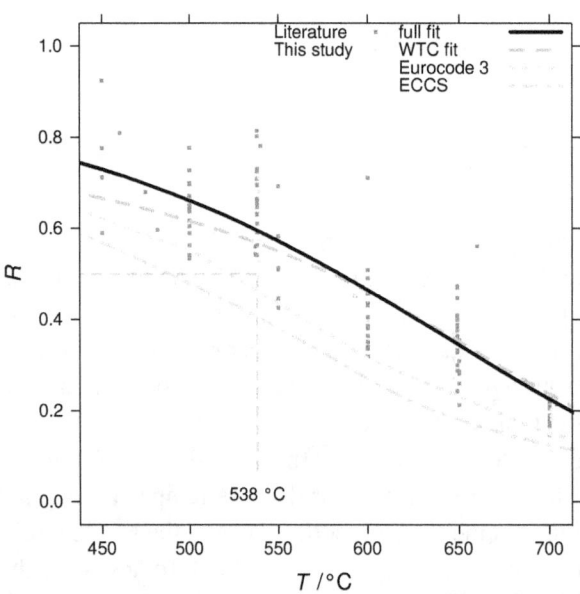

(b) expanded temperature range

Figure 50: ECCS recommended effective retained yield strength compared to other retained strength models. (a) full temperature range. (b) temperature range where strength drop is largest.

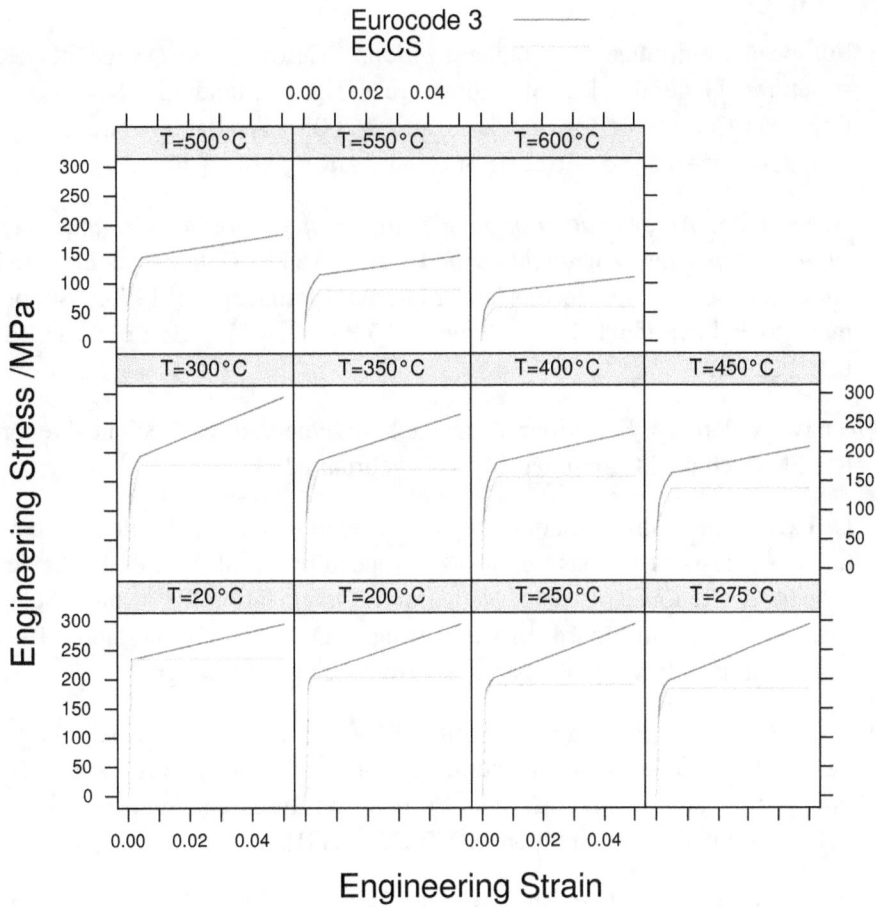

Figure 51: ECCS stress-strain model compared to the Eurocode 3 stress-strain model.

from an analytical function, but are instead tabulated as individual points at specified temperatures for different grades of steel. Consistent with the retained strength behavior, the ECCS model predicts lower strength for all strains and temperatures.

References

[1] European Committee for Standardization. Eurocode 3. Design of steel structures. General rules. Structural fire design. Standard EN 1993-1-2, European Committee for Standardization, 2005. Available from: `http://eurocodes.jrc.ec.europa.eu/showpage.php?id=133`.

[2] G. V. Smith. *An evaluation of the elevated temperature, tensile, and creep-rupture properties of wrought carbon steel*. ASTM Data Series DS 11S1. American Society for Testing and Materials, Philadephia, 1970. Supplement to Publication DS 11, formerly STP 180 [65]. `doi:10.1520/DS11S1-EB`.

[3] Yngve Anderberg. Behaviour of steel at high temperatures. Technical report, RILEM Technical Committee 44-PHT, February 1983.

[4] L. Twilt. Stress-strain relationsships of structural steel at elevated temperatures: Analysis of various options & european proposal, Part F: mechanical properties. Report BI-91-015, Netherlands Organization for Applied Scientific Research, Van Mourik Broekmanweg 6, Delft, The Netherlands, 1991. The report is only available by request from `http://www.tno.nl`.

[5] V. Jerath, K. J. Cole, and C. I . Smith. Elevated temperature tensile properties of structural steels manufactured by the British Steel Corporation. Report T/RS/1189/80/C, British Steel Corporation Research Organization, July 1980. document has annotation ISO TC22 WG318.

[6] A. G. Stevens, R. C. Cornish, and D. H. Skinner. Tensile data on four structural steels. Technical Report MRL 6/5, Melbourne Research Laboratories, The Broken Hill Proprietary Company, Clayton, Vic. Australia, November 1971. also cited in 33.

[7] Yngve Anderberg. Modelling steel behaviour. *Fire Safety Journal*, 13(1):17–26, 1988. `doi:10.1016/0379-7112(88)90029-X`.

[8] G.M.E. Cooke. An introduction to the mechanical properties of structural steel at elevated temperatures. *Fire Safety Journal*, 13(1):45–54, 1988. `doi:10.1016/0379-7112(88)90032-X`.

[9] ECCS-Technical Committee 3-Fire Safety of Steel Structures. European recommendations for the fire safety of steel structures: Calculation of the fire resistance of load bearing elements and structural assemblies exposed to the standard fire. Technical report, European Con-

128

vention for Constructional Steelwork, Brussels, Belgium, 1983. Available from: `http://www.eccspublications.eu/index.php?section=library&content=&act=detail&id=95`.

[10] ASTM International. Standard test methods for tension testing of metallic materials [metric]. Standard E8/E8M-09, ASTM International, W. Conshohocken, Pa, 2002. `doi:10.1520/E0008_E0008M-09`.

[11] ASTM International. Standard test methods for elevated temperature tension tests of metallic materials. Standard E21-09, ASTM International, W. Conshohocken, Pa, 2009. `doi:10.1520/E0021-09`.

[12] William E. Luecke, J. David McColskey, Christopher N. McCowan, Stephen W. Banovic, Richard J. Fields, Timothy Foecke, Thomas A. Siewert, and Frank W. Gayle. Federal building and fire safety investigation of the World Trade Center disaster: Mechanical properties of structural steel. Technical Report NCSTAR 1-3D, National Institute of Standards and Technology, 2005. Available on-line at http://wtc.nist.gov. Available from: `http://wtc.nist.gov/NCSTAR1/PDF/NCSTAR%201-3D%20Mechanical%20Properties.pdf`.

[13] T. Z. Harmathy and W. W. Stanzak. Elevated-temperature tensile and creep properties of some structural and prestressing steels. In *Fire Test Performance*, number STP 464, pages 186–208. American Society for Testing and Materials, 1970. `doi:10.1520/STP44718S`.

[14] Jörgen Thor. Effect of creep on load-bearing capacity of steel beams exposed to fire. Publication 24, Swedish Institute of Steel Construction, Stockholm, Sweden, 1971. in English.

[15] T. T. Lie and W. W. Stanzak. Empirical method for calculating fire resistance of protected steel columns. *Engineering Journal, Transactions of the Canadian Society for Civil Engineering*, 57(5/6):73–80, 1974. Research Paper No. 619 of the Division of Building Research, National Research Council of Canada. Available from: `http://irc.nrc-cnrc.gc.ca/pubs/rp/rp574/rp619.pdf`.

[16] M. Holmes, R. D. Anchor, G. M. E. Cook, and R. N. Crook. The effects of elevated temperatures on the strength properties of reinforcing and prestressing steels. *The Structural Engineer*, 60B(1):7–13, 1982.

[17] H. L. Malhotra. Report on the work of technical committee 44-PHT "properties of materials at high temperatures". *Materials and Structures*, 15:161–170, 1982. doi:10.1007/BF02473577.

[18] B.R. Kirby and R.R. Preston. High temperature properties of hot-rolled, structural steels for use in fire engineering design studies. *Fire Safety Journal*, 13(1):27–37, 1988. doi:10.1016/0379-7112(88)90030-6.

[19] Jyri Outinen, Jyrki Kesti, and Pentti Mäkeläinen. Fire design model for structural steel S355 based upon transient state tensile test results. *Journal of Constructional Steel Research*, 42(3):161 – 169, 1997. doi:10.1016/S0143-974X(97)00018-7.

[20] Pentti Mäkeläinen, Jyri Outinen, and Jyrki Kesti. Fire design model for structural steel S420M based upon transient-state tensile test results. *Journal of Constructional Steel Research*, 48(1):47–57, 1998. doi:10.1016/S0143-974X(98)00005-4.

[21] G. Q. Li, S. C. Jiang, Y. Z. Yin, K. Chen, and M. F. Li. Experimental studies on the properties of constructional steel at elevated temperatures. *J. Struct. Engineering-ASCE*, 129:1717–1721, 2003. doi:10.1061/(ASCE)0733-9445(2003)129:12(1717).

[22] Yasushi Mizutani, Kenichi Yoshii, Rikio Chijiiwa, Kiyoshi Ishbashi, Yoshiuki Watanabe, and Uzuru Yoshida. 590MPa class fire-resistant steel for building structural use. Technical report 90, Nippon Steel, July 2004. UDC 669.14.108.292:691.714. Available from: http://www.nsc.co.jp/en/tech/report/no90.html.

[23] Ju Chen, Ben Young, and Brian Uy. Behavior of high strength structural steel at elevated temperatures. *J. Struct. Eng. ASCE*, 132(12):1948–1954, 2006. doi:10.1061/(ASCE)0733-9445(2006)132:12(1948).

[24] Rikio Chijiiwa, Hiroshi Tamehrio, Yuzuru Yoshida, Kazuo Funato, Tyuji Uemori, and Yukihiko Horii. Development and practical application of fire-resistant steel for buildings. Nippon Steel Technical Report 58, Nippon Steel Corporation, July 1993. Special Issue on New Steel Plate Products of High Quality and High Performance UDC669.14.018.291 : 699.81. Available from: http://www.nsc.co.jp/en/tech/report/no58.html.

[25] B. C. Gowda. Tensile properties of SA516, grade 55 steel in the temperature range of 25 °C–927 °C and strain rate range of 10^{-4} to 10^{-1} sec^{-1}. In

George V. Smith, editor, *Characterization of Materials for Service at Elevated Temperatures*, pages 145–158, New York, 1978. The American Society of Mechanical Engineers. presented at the 1978 ASME/CSME Montreal Pressure Vessel & Piping Conference, Montreal, Quebec, Canada, June 25–29 1978.

[26] J. M. Holt. Short-time elevated-temperature tensile properties of USS Cor-Ten and USS Tri-Ten high-strength low-alloy steels, USS Man-Ten (A 440) high-strength steel, and ASTM A 36 steel. Progress Report 57.19-901(1), Applied Research Laboratory, United States Steel Corporation, Monroeville, Pa, 1964. Data in Figure 1.5 appears in 26 and 27.

[27] J. M. Holt. Short-time elevated-temperature tensile properties of USS "T-1" and USS "T-1" type A constructional alloy steels. Progress Report 57.19-901(3)(a-AS-EA-2, Applied Research Laboratory, United States Steel Corporation, Monroeville, Pa, date unknown.

[28] B. R. Kirby and R. R. Preston. The application of BS5950:Part 8 on fire limit state design to the performance of 'old' structural mild steels. Report SL/RS/R/S1199/17/91/C, British Steel, Swinden Laboratories, Moorgate, Rotherham,S60 3AR, April 1991. Available from: `http://www.mace.manchester.ac.uk/project/research/structures/strucfire/DataBase/References/ApplicationBS5950.pdf`.

[29] S. Lou and D. O. Northwood. Effect of temperature on the lower yield strength and static strain ageing in low-carbon steels. *Journal of Materials Science*, 30:1434–1438, 1995. `doi:10.1007/BF00375244`.

[30] Jyri Outinen, Olli Kaitila, and Pentti Mäkeläinen. High-temperature testing of structural steel and modelling of structures at fire temperatures. Report TKK-TER-23, Helsinki University of Technology, Laboratory of Steel Structures, 2001. Available from: `http://www.tkk.fi/Units/Civil/Steel/Publications/TKK_TER_series/TKK-TER-23.pdf`.

[31] Kok Weng Poh. *Behaviour of Load-Bearing Members in Fire*. PhD thesis, Dept. of Civil Engineering, Monash University, Clayton, Victoria, Australia, 1998.

[32] Y. Sakumoto. Research on new fire-protection materials and fire-safe design. *J. Struct. Engineering-ASCE*, 125(12):1415–1422, 1999. `doi:10.1061/(ASCE)0733-9445(1999)125:12(1415)`.

[33] D. H. Skinner. Steel properties for prediction of structural performance during fires. In Peter Swannell, editor, *Fourth Australasian conference on the mechanics of structures and materials: Proceedings*, pages 269–276, 1973. Papers presented at the 4th conference held on August 20th-22nd, 1973 at University of Queensland, Brisbane, Queensland.

[34] C. E. Spaeder Jr. and R. M. Brown. Elevated temperature characteristics of Cor-Ten high-strength low-alloy steels. In George V. Smith, editor, *Effects of melting and processing variables on the mechanical properties of steel*, number MPC-6, pages 273–307, New York, 1977. American Society of Mechanical Engineers. presented at the winter annual meeting of the American Society of Mechanical Engineers, Atlanta, Georgia, November 27-December 2, 1977 ; sponsored by the Metal Properties Council, Inc. the Materials Division, ASME.

[35] United State Steel Corporation. Steels for elevated temperature service. Technical report, United State Steel Corporation, 1972.

[36] International Organization for Standardization. Metallic materials–tensile testing at elevated temperature. Standard 783:1999, International Organization for Standardization, 1999. superseded by ISO 6892-2:2011. Available from: http://www.iso.org/iso/iso_catalogue/catalogue_tc/catalogue_detail.htm?csnumber=26858.

[37] Japan Industrial Standards. Method of elevated temperature tensile test for steels and heat-resisting alloys. Standard JIS G 5067:1998, Japan Industrial Standards, 1998. Available from: http://www.webstore.jsa.or.jp/webstore/JIS/FlowControl.jsp.

[38] G. Williams-Leir. Creep of structural-steel in fire - analytical expressions. *Fire and Materials*, 7:73–78, 1983. doi:10.1002/fam.810070205.

[39] L. Twilt. Strength and deformation properties of steel at elevated temperatures: Some practical implications. *Fire Safety Journal*, 13(1):9–15, 1988. doi:10.1016/0379-7112(88)90028-8.

[40] Guanyu Hu, Mohammed Ali Morovat, Jinwoo Lee, Eric Schell, and Michael Engelhardt. Elevated temperature properties of astm a992 steel. In Mark Waggoner Lawrence G. Griffis, Todd Helwig and Marc Holt, editors, *Proceedings of the Structures Congress 2009 Don't Mess with Structural Engineers–Expanding our Role*. American Society of Civil Engineers, 2009. doi:10.1061/41031(341)118.

[41] H. J. Frost and M. F. Ashby. *Deformation-mechanism maps: the plasticity and creep of metals and ceramics.* Pergamon Press, 1982. Available on-line at urlhttp://engineering.dartmouth.edu/defmech/.

[42] S. Takeuchi and A. S. Argon. Steady-state creep of single-phase crystalline matter at high temperature. *Journal of Materials Science,* 11:1542–1566, 1976. doi:10.1007/BF00540888.

[43] G. R. Johnson and W. H. Cook. A constitutive model and data for metals subjected to large strains, high strain rates and high temperatures. In *Proceedings, Seventh International Symposium on Ballistics, the Hague, the Netherlands, 19-21 April, 1983,* pages 541–547. American Defense Preparedness Association and Koninklijk Instituut van Ingenieurs, 1983.

[44] F. J. Zerilli and R. W. Armstrong. Dislocation-mechanics-based constitutive relations for material dynamics calculations. *J. Appl. Phys.,* 61:1816–1825, 1987. doi:10.1063/1.338024.

[45] K. W. Poh. Stress-strain-temperature relationship for structural steel. *Journal of Materials in Civil Engineering,* 13(5):371–379, SEP-OCT 2001. doi:10.1061/(ASCE)0899-1561(2001)13:5(371).

[46] ASTM International. Standard terminology relating to methods of mechanical testing. Standard Terminology E6-09b, ASTM International, W. Conshohocken, Pa, 2009. doi:10.1520/E0006-09B.

[47] American Institute of Steel Construction. *Steel Construction Manual.* American Institute of Steel Construction, 14 edition, 2011.

[48] National Institute of Standards and Technology. *NIST/SEMATECH e-Handbook of Statistical Methods.* National Institute of Standards and Technology, 2010. Available from: http://www.itl.nist.gov/div898/handbook/pmd/section4/pmd431.htm.

[49] ASTM International. Standard test method for optical emission vacuum spectrometric analysis of carbon and low-alloy steel. Standard E415-08, ASTM International, W. Conshohocken, Pa, 2008. doi:10.1520/E0415-08.

[50] ASTM International. Standard test methods for determination of carbon, sulfur, nitrogen, and oxygen in steel and in iron, nickel, and cobalt alloys. Standard E1019-03, ASTM International, W. Conshohocken, Pa, 2003. doi:10.1520/E1019-08.

[51] William E. Luecke. Federal building and fire safety investigation of the World Trade Center disaster: Contemporaneous structural steel specifications. Technical Report NCSTAR 1-3A, National Institute of Standards and Technology, 2005. Available on-line at http://wtc.nist.gov.

[52] R Development Core Team. *R: A Language and Environment for Statistical Computing*. R Foundation for Statistical Computing, Vienna, Austria, 2010. ISBN 3-900051-07-0. Available from: `http://www.R-project.org`.

[53] J. M. Robinson and M. P. Shaw. Microstructural and mechanical influences on dynamic strain aging phenomena. *International Materials Reviews*, 39(3):113–122, 1994. Available from: `http://www.ingentaconnect.com/content/maney/imr/1994/00000039/00000003/art00002`.

[54] B.J Brindley and J.T Barnby. Dynamic strain ageing in mild steel. *Acta Metallurgica*, 14(12):1765 – 1780, 1966. `doi:10.1016/0001-6160(66)90028-9`.

[55] T. Borvik, O. S. Hopperstad, S. Dey, E. Pizzinato, M. Langseth, and C. Albertini. Strength and ductility of Weldox 460 E steel at high strain rates, elevated temperatures and various stress triaxialities. *Engineering Fracture Mechanics*, 72(7):1071–1087, 2005. `doi:10.1016/j.engfracmech.2004.07.007`.

[56] Morihisa Fujimoto, Fukujiro Furumura, Takeo Ave, and Yasuji Shinohara. Primary creep of structural steel (SS41) at high temperatures. *Transactions of the Architectural Institute of Japan*, 296:145–157, 1980. Nihon Kenchiku Gakkai ronbun hokokushu. Available from: `http://ci.nii.ac.jp/naid/110003881722/en`.

[57] G. F. Melloy and J. D. Dennison. Short-time elevated-temperature properties of A7, A440, A441 V50, and V65 grades. Inter-office correspondence 1900-4g, Bethlehem Steel, 1963.

[58] R. L. Brockenbrough and B. G. Johnston. Steel design manual. Technical Report ADUSS 27-3400-01, United States Steel Corporation, July 1968. Some data appear in Holt [26].

[59] Canadian Institute of Steel Construction. Obsolete canadian structural steel grades, 1935–1971. Technical report, Canadian Institute of Steel Construction, Markham, ON, Canada, 2005. Available from: `http://www.cisc-icca.ca/resources/tech/historical/grades/`.

134

[60] B.R. Kirby. The application of BS5950: Part 8 of fire limit state design to the performance of 'old' structural mild steels. *Fire Safety Journal*, 20(4):353–376, 1993. Also published as British Steel Technical report SL/RS/R/S1199/17/91/C [28]. doi:10.1016/0379-7112(93)90055-U.

[61] S. Lou and D. Northwood. Effect of strain aging on the strength coefficient and strain-hardening exponent of construction-grade steels. *Journal of Materials Engineering and Performance*, 3:344–349, 1994. doi:10.1007/BF02645330.

[62] M. J. Manjoine. Influence of rate of strain and temperature on yield stress of mild steel. *Transactions of the American Society of Mechanical Engineers*, 66:A–211–A–218, 1944.

[63] S. Kotwal. The evolution of australian material standards for structural steel. *Steel Construction, Journal of the Australian Steel Institute*, 33(2):3–19, 1999. Available from: http://www.steel.org.au/inside.asp?ID=453&pnav=401.

[64] Koichi Takanashi. SN steel: Performances required for steel proudcts for building construction. *Steel Construction Today and Tomorrow*, (1):4–7, 2003.

[65] Ward F. Simmons and Howard Clinton Cross. *Elevated-temperature properties of carbon steels*. Number 180 in ASTM Special Technical Publication. American Society for Testing Materials, Philadephia, Pa, 1955. doi:10.1520/STP180-EB.

[66] S. H. Ingberg and P. D. Sale. Compressive strength and deformation of structural steel and cast-iron shapes at temperatures up to 950 °C. (1742° F). *Proceedings of the American Society for Testing Materials*, 26:33–55, 1926.

[67] G. Verse. Elastic properties of steel at high temperatures. *American Society of Mechanical Engineers – Transactions*, 57(1, sec 1):1 – 4, 1935.

[68] ASME. Boiler and pressure vessel code. section II materials, part D properties (customary), subpart 2 physical properties tables. Code, ASME International, New York, 2004. Table TM-1.

[69] C.L Clark. *High-temperature Alloys*. Pitman, New York, 1953.

[70] E.H.F. Date. Elastic Properties Of Steels. *Journal of the Iron and Steel Institute*, 207(Part 7):988–991, 1969.

[71] F. Garofalo. The influence of temperature on the elastic constants of some commercial steels. In *Symposium on Determination of Elastic Constants*, number STP 129. American Society for Testing and Materials, 1952. presented at the fifty-fifth annual meeting (fiftieth anniversary meeting) American Society for Testing Materials, New York, N.Y., June 25, 1952.

[72] F. Garofalo. Survey of various special tests used to determine elastic, plastic, and rupture properties of metals at elevated temperatures. *Journal of Basic Engineering*, pages 867–881, December 1960. Paper 59-A-112.

[73] W. Köster. Die Temperaturabhängigkeit des Elastizitätsmoduls reiner Metalle. *Z. Metallkd*, 39(1):1–9, 1948.

[74] J.C. Ritter and McPherson R. Anisothermal Stress Relaxation in a Carbon-Manganese Steel. *Journal of the Iron and Steel Institute*, 208:935–941, 1970.

[75] M.H. Roberts and J. Nortcliffe. Measurement of Young modulus At High Temperatures. *Journal of the Iron And Steel Institute*, 157(3):345–348, 1947.

[76] T. Uddin and C. G. Culver. Effects of elevated temperature on structural members. *J. Struct. Div. ASCE*, 101(7):1531–1549, 1975. data for Tall, Stanzak.

[77] T. T. Lie, editor. *Structural Fire Protection: Manual of Practice*. ASCE Manuals and Reports of Engineering Practice; no. 78. American Society of Civil Engineers, 1992.

[78] Venkatesh Kodur, Mahmud Dwaikat, and Rustin Fike. High-temperature properties of steel for fire resistance modeling of structures. *Journal of Materials in Civil Engineering*, 22(5):423–434, 2010. doi:10.1061/(ASCE)MT.1943-5533.0000041.

[79] Julian R. Frederick. Ultrasonic measurement of the elastic properties of polycrystalline materials at high and low temperatures. *Journal of the Acoustical Society of America*, 20(4):586–586, 1948. Program of the Thirty-Fifth Meeting of the Acoustical Society of America. doi:10.1121/1.1916961.